BEYOND AVIATION HUMAN FACTORS

Beyond Aviation Human Factors

Safety in High Technology Systems

**Daniel E. Maurino   James Reason
Neil Johnston   Rob B. Lee**

ASHGATE

Published by
Ashgate Publishing Limited
Gower House
Croft Road
Aldershot
Hants GU11 3HR
England

Ashgate Publishing Company
131 Main Street
Burlington, VT 05401-5600 USA

Hardback edition reprinted 1997, 1998, 2000
Paperback edition edition published in 1998
Paperback edition reprinted 2000, 2001 [twice], 2002

Ashgate website: http://www.ashgate.com

British Library Cataloguing in Publication Data
Maurino, Dan
    Beyond Aviation Human Factors
    I. Title
    629.13

Library of Congress Catalog Card Number: 95-75482

ISBN 0-291-39822-7 [Hbk]
ISBN 1-84014-948-5 [Pbk]

Printed and bound in Great Britain by M.P.G. Books Ltd Bodmin, Cornwall

Contents

Foreword

The Honourable Mr. Justice Virgil P. Moshansky

The authors of this book have focused upon and captured with remarkable clarity the essence of the investigation conducted by the Commission of Inquiry which I was privileged to head, into the March 10, 1989 Air Ontario F-28 jet crash at Dryden, Ontario.

I am pleased to have been invited by the authors to write a Foreword to this book, briefly touching upon some aspects of the Dryden Commission and providing some insight into a few of the behind-the-scenes events which occurred during the three year life of the Commission.

In retrospect, two fortuitous decisions made at the outset contributed greatly to whatever success the Commission achieved. The first was my insistence upon the transfer of responsibility for the Commission from the office of the Minister of Transport, where it initially resided, to that of the Privy Council. It was simply unacceptable for the Commission to report to a Minister whose department's role I anticipated possibly having to investigate. The other was the insistence on the provision by the Government of a broad mandate in the Commission's terms of reference, in order to permit an unbridled inquiry into the entire aviation system, in a search for systemic failures contributing to the crash at Dryden. Both requests were acceded to by the Government.

This assured financial and moral support for the Inquiry at the highest levels of Government, and provided complete independence from Transport Canada, the area of Government most likely to be investigated for possible failures in the discharge of its responsibility for aviation safety. Vindication for this approach manifested itself almost immediately after my appointment. Early requests by my team of accident investigators for disclosure of pertinent Transport Canada documentation and records were not looked upon kindly by senior officials and counsel in the department. It was only upon threat of the issue of witness subpoenas to the most senior management officials of the regulator requiring their attendance before the Inquiry, and their production of relevant documents, that the records in question were eventually made available to my Commission.

The crash at Dryden occurred on the heels of the release of a highly controversial Canadian Aviation Safety Board (CASB) Report into that agency's investigation of the December 15, 1985 Arrow Air DC-8 crash at Gander, Newfoundland.

To address widespread and vocal public concern in Canada over its handling of the Gander crash, that investigative body was legislated out of existence by the Government of Canada on March 29, 1989. Concurrently, on that date my Commission of Inquiry was established under the *Inquiries Act* to take over the investigation of the Dryden crash from the now defunct CASB.

Having been appointed as Commissioner in the midst of the Gander controversy, I was conscious of the need for my Commission to be not only totally independent in fact, but also to be so perceived by the public. No less than the restoration of credibility to the aviation accident investigation process in Canada was involved.

The physical cause of the Dryden crash, i.e. the contamination of the aircraft surfaces with ice and snow, prior to take-off, was evident within days of the commencement of the Inquiry. It would have been a simple matter to assign the blame to pilot error and to let it go at that, as indeed had been done in approximately 80% of aviation accident investigations in the past. However, the Commission represented a rare opportunity to examine the entire aviation system for latent and active failures which might have contributed to the Captain's faulty decision. A conscious decision was made to launch an in-depth search for such failures, and to investigate fully the impact of Human Factors throughout the aviation system upon the events at Dryden.

During the early stages of the Inquiry, counsel for the regulator attempted to limit the scope of the Inquiry with threats to limit my mandate by seeking an order in the Federal Court of Canada. When it became clear that intimidation would not succeed, these attempts were abandoned and some senior management personnel changes occurred within Transport Canada, leading to a genuine spirit of cooperation with the Inquiry. To Transport Canada's credit, the Dryden Inquiry was subsequently embraced by the regulator as a means of achieving a long overdue cleansing and restructuring of the Canadian aviation system.

There is no question that were it not for the wide powers enjoyed by a Commissioner under the *Inquiries Act,* including the power to subpoena witnesses to testify under oath and for the production of documents, the far-ranging investigation which was undertaken into the Dryden crash would not have otherwise been possible. While a Commission of Inquiry is clearly not a practical way to conduct all aviation accident inquiries, such a Commission, when established to investigate a major aviation accident, is the most powerful and effective instrument available for an in-depth examination of the safety of a nation's aviation system.

In the case of the regulator, the work of my Commission revealed serious policy lapses, inconsistent policies, bureaucratic bungling and human failures at all levels of Government and the bureaucracy which contributed to the events leading to Dryden. On the plus side, the work of the Commission has resulted in major structural and policy changes within Transport Canada. To its great credit the regulator, even while the Inquiry was ongoing, and in the three years which have elapsed since the release of my Final Report and its 191 recommendations, has moved expeditiously to make the changes recommended. Incompetents in senior positions of responsibility have been replaced by highly qualified and dedicated individuals. They have organized and managed an unprecedented cooperative effort involving representatives of the aviation industry, professional pilots and regulators in the implementation of the recommendations of my Commission.

Insofar as the carrier was concerned, there were also numerous attempts through its counsel to limit the scope of the Inquiry, accompanied by threats from time to time to take this issue to the Federal Court. During the course of the writing of the Final Report, an action was brought in the Federal Court of Canada to prevent me from naming in my Report any person whose human failures or shortcomings might be found to have contributed in any way to the events which led to the tragedy at Dryden. The action was successfully resisted, but it added several months to the life of the Commission and necessitated countless hours of additional work by my staff and myself, and the expenditure of considerable funds. In the end, those who were implicated were named.

It is commonplace in all aspects of life to find resistance to change. This truism certainly manifested itself during the Dryden Inquiry. There was initial resistance to change on the part of the regulator and ongoing resistance on the part of the carrier. To the very end the

President and CEO of the carrier refused to admit that the Dryden accident was a preventable accident. All the international aviation experts who testified at the Inquiry were unanimous in stating that all ground icing accidents are preventable.

Even within the ranks of my Commission officials there was a widespread initial reluctance to pursue what I believe became one of the two most important aspects of my Commission; the exhaustive investigation of the subject of aircraft ground de-icing and anti-icing, procedures and fluids, which I decided to undertake in the interests of aviation safety. I had been advised that the subject should not be pursued. Some argued that it was beyond the scope of my mandate. Others expressed a preference not to become involved in the philosophical difference of opinion which existed between the North American and the European carriers and regulators on the subject of aircraft ground icing and the use of Type II anti-icing fluids. However, once the decision was made to proceed with a comprehensive investigation into the subject, all of my Commission officials undertook the challenge with enthusiasm. It is gratifying to note the profound changes in aircraft ground de-icing and anti-icing methods and procedures which have occurred as a result of the investigation which took place and the recommendations emerging from it.

The plethora of human factors within the aviation system which were demonstrated to have influenced the events at Dryden represent the other of the two most important subjects examined in-depth by the Dryden Commission; a subject which of course forms the genesis of this book.

I applaud the authors of this work for undertaking this worthy international endeavour. Having had the opportunity of reading its chapters in their draft stage, I am confident that this book will earn for them the admiration and respect of the aviation community worldwide.

Preface

This is a book about Human Factors, and yet it is not. It discusses the relentless threat to aviation safety as well as the central preoccupation of aviation Human Factors: human error. Therefore, by loose standards, it could be included among the abundant literature on aviation Human Factors produced over the last few years. This book, however, discusses human error from a different perspective when compared with the views on human error favoured by 'traditional' Human Factors. Such views, essentially fostered by clinical and behavioural psychology, and to a degree by the engineering sciences, address human error focusing on the individual human operator – mainly pilots, controllers or mechanics. The mechanisms of individual human error are sought to be understood, so that measures to minimize the possibility of its appearance can be devised. The damaging consequences of the quota of error which will inevitably permeate even the best erected defences are sought to be minimized. In real life terms, one consequence of these views has been the rather naïve branding of pilots, controllers and mechanics as the sole professions responsible for safety in aviation.

This book proposes an alternative approach: the proactive management of human error from an organizational viewpoint. The authors believe that the mechanisms of individual human error are by now well understood, and defences to cope with its known damaging consequences are in place. Therefore, when thinking about aviation safety and effectiveness, there is little to be gained by further pursuing individual-oriented avenues of action. In the best of the scenarios, the profit will not be commensurate with the investment of resources. The authors are convinced that the aviation industry is at the foothills of significant change in approaches to pursue system safety, and that such approaches must necessarily take a proactive rather than a reactive stance. Furthermore, these approaches must be based in the very obvious notion that human error does not take place in a vacuum, but within the context of organizations which either resist or foster it.

The central contention of this book is simple and straightforward: no matter how well equipment is designed, no matter how sensible regulations are, no matter how much humans excel in their individual or small team's performance, they can never be better than the system which bounds them. It is time to start thinking about aviation safety in collective rather than individual terms. Therefore, in advocating an organizational, systemic approach to human error reduction, this book goes several steps beyond aviation Human Factors.

This book is intended as a bridge between academic knowledge and aviation practice. The academic community has, for over twenty years, researched system approaches to safety in high-technology systems, a perfect example of which is aviation. This research, however, is only starting to be applied to practice, and there is a long way to go before an effective transfer of knowledge takes place, at least in aviation. The importance of dropping the piecemeal approaches which have plagued past safety endeavours in favour of proactive, systemic and organizational approaches is essential. This change must be preceded by an educational campaign to explain *why* change is necessary. The International Civil Aviation Organization (ICAO) is spearheading this campaign. This book is yet another building block in it. Given the experience and qualifications of the intended readership, every effort has been made to advance the academic knowledge in a clear and simple manner. By resorting to case studies, the relevance of the knowledge – fundamentally in terms of prevention and, to a lesser degree, in terms of investigation – to the operational environment becomes crystal-clear.

The organizational framework pioneered by Professor James Reason for the proactive analysis of safety in high-technology systems is at the heart of this book. Such a framework has an enduring role to play, both now and in the immediate future. Within the aviation industry, this analytic framework has been adopted by ICAO, the International Air Transport Association (IATA), the International Federation of Air Traffic Controllers Associations (IFATCA), the National Transportation Safety Board (NTSB) of the United States, the Bureau of Air Safety Investigations (BASI) of Australia, the Transportation Safety Board (TSB) of Canada and British Airways. Hopefully, more will follow.

Throughout its almost one hundred years of history, different periods in aviation favoured different approaches to the control and avoidance of human error. These included widely ranging strategies, varying from exhortations to professional behaviours in one extreme, to the attempt to displace humans from control through large-scale automation and technology at the other, with numerous combinations in between. Moreover, at each opportunity, the approach of preference was heralded by its proponents as the final solution to human error in aviation. Human Factors itself could not escape from such misleading simplification, having once been proclaimed – some twenty years ago – the last frontier in aviation safety. Obviously, it is not.

There is no challenge to the importance of Human Factors, nor to the fact that aviation safety endeavours have been ribboned by unparalleled success when compared to other transportation or production industries. However, those advocating some final solution of sorts to human error – or indeed, contending that they have found one – ignore the painful evidence provided by a trail of smouldering wreckage. Such a final solution does not exist. Pursuing safety and effectiveness in aviation is an endless quest, which allows for no time to rest over accomplishments. It requires particular solutions to specific deficiencies which present themselves under substantially different symptoms and circumstances over time. The approach and solutions proposed by this book are no exceptions to this. They are by no means a last frontier. But its judicious combination with existing strategies of proven worth presents – in the view of the authors – a contemporary avenue to progress safety and effectiveness of civil aviation until the passage of time dictates otherwise.

1 Widening the search for accident causes: a theoretical framework

Introduction

This book is about extending the scope of accident analysis from individuals to organizations, from the 'sharp end' to the top-level management of the air transportation system as a whole. In aviation, as in other complex technologies, we are in the age of the organizational accident (Reason, 1990). That is, accidents in which pre-existing and often long-standing latent failures, arising in the organizational and managerial sectors, combine with local triggering conditions – on the flight deck, in air traffic control centres and in maintenance facilities – to penetrate or bypass the aviation system's multiple defences.

Every age has a dawning. In the maritime world, Mr Justice Sheen's judgement on the causes of the capsize of the *Herald of Free Enterprise* represents one such marker. After acknowledging the shipboard (or active) errors of the Master, the Chief Officer and the Assistant Bosun, he wrote:

> But a full investigation into the disaster leads inexorably to the conclusion that the underlying or cardinal faults lay higher up in the Company . . . From top to bottom the body corporate was infected with the disease of sloppiness. (Sheen, 1987)

In aviation, we are indebted to Commissioner Moshansky's inquiry into the crash at Dryden, Ontario, perhaps the most exhaustive investigation ever undertaken into an aircraft accident. On the face of it, the accident was due to the Captain's flawed decision to take off in a heavy snow squall without de-icing protection from ground contamination. But Commissioner Moshansky interpreted his brief more widely. Introducing his findings, he wrote:

> The accident at Dryden on March 10, 1989, was not the result of one cause but of a combination of several related factors. Had the system operated effectively, each of the factors might have been identified and corrected before it took on significance. It will be shown that this accident was the result of a failure in the air transportation system. (Moshansky, 1992)

The main purpose of this chapter is to outline a theoretical framework that seeks to provide a principled basis both for understanding the causes of organizational accidents and for creating a practical remedial toolbag that will minimize their occurrence. This framework traces the development of an accident sequence from organizational and managerial decisions, to conditions in various workplaces (flight decks, hangars, etc.), and thence to the personal and situational factors leading to errors and violations. It identifies both an active and a latent failure pathway to an event, where an event is defined as the breaching, absence or bypassing of some or all of the system's various defences and safeguards. Such an event may have disastrous consequences or it may merely serve as a 'free lesson'. The outcome depends upon the local circumstances and how many of the defences-in-depth are removed.

The pursuit of safety in aviation, as elsewhere, has seen many fashions. Is this organizational approach just another passing fad, or does it have something of real substance to offer? To answer this question, we have to look in some detail at the tangled question of individual versus collective (i.e. managerial and organizational) contributions to accidents.

Individual or collective errors?

This issue has a number of related dimensions. The first is a moral one, relating to blame, responsibility and legal liability. The second is scientific, having to do with the nature of cause and effect in an accident sequence. The third is entirely practical, and concerns which standpoint, individual or collective, leads to more effective counter-measures.

The moral dimension

From a moral or legal perspective, there is much to be gained from pursuing an individual rather than a collective approach to accident causation. The reasons are listed below.

- It is much easier to pin the legal responsibility for an accident upon the errors and violations of those in direct control of the aircraft or vessel at the time of the accident. The connection between these individual actions and the disastrous outcome is far more easily proved than are any possible links between earlier management decisions and the accident. This was clearly shown by the failed prosecution of the managers implicated (by the Sheen Inquiry) in the capsize of the *Herald of Free Enterprise.*
- This is further compounded by the willingness of professionals such as aircraft captains and ships' masters to accept this responsibility. They are accorded substantial authority, power and prestige and, in return, they are expected (and expect) to 'carry the can' when things go wrong. The buck and the blame traditionally stops with them.
- Most people place a large value on personal autonomy, or the sense of free will. We also impute this to others, so that when we learn that someone has committed an error with bad consequences, we assume that this individual actually *chose* an error-prone rather than a 'sensible' course of action. In other words, we tend to perceive the errors of others as having an intentional element, particularly when their training and status suggest that 'they should have known better'. Such voluntary actions attract blame and recrimination, which in turn are felt to deserve various sanctions.
- Our judgements of human actions are subject to similarity bias. We have a natural tendency to assume that disastrous outcomes are caused by equally monstrous blunders. In reality, of course, the magnitude of the disaster is determined more by situational factors than by the extent of the errors. Many bad accidents rise from a concatenation of relatively minor failings in different parts of the system (e.g. the Tenerife runway disaster).
- Finally, it cannot be denied that there is a great deal of emotional satisfaction to be gained from having someone (rather than something) to blame when things go badly wrong. Few of us are able to resist the pleasures of venting our psychic spleens on some convenient scapegoat. And in the case of organizations, of course, there is considerable financial advantage in being able to

3

detach individual fallibility from corporate responsibility.

The scientific dimension

Should one halt the search for causes after identifying the human and/or component failures immediately responsible for the accident (as has been done in many accident investigations), or is it scientifically more appropriate to track back to their organizational root causes? On the face of it, the answer seems obvious. Yes, it must be better (in the sense of being a more accurate representation of the true state of affairs) to try to find all the systemic factors responsible for the accident. But the issue is not quite so simple. Let us examine some of the problems.

- Why should we stop at the organizational roots? In a deterministic world, everything has a prior cause. In theory, therefore, we could go back to the Big Bang. Seen from this broader historical perspective, an analytical stop-rule located at the organizational root causes is just as arbitrary – in strict scientific terms – as one located close to the proximal individual failures.
- There is a way out of this muddle, but it is more practical than scientific. In seeking the reasons for an accident, we should go far enough back to identify causal factors that, if corrected, would enhance the system's resistance to subsequent challenges. The people most concerned and best equipped to do this are those within the organization(s) involved, so it makes practical sense to stop the analysis at these organizational boundaries. However, such boundaries are often indistinct, particularly in aviation where there are a large number of inter-related sub-systems involved.
- Perhaps the most serious scientific problem, however, has to do with the peculiar nature of accidents and with the way in which they change our perceptions of preceding events. In retrospect, an accident appears to be the point of convergence of a number of causal chains. Looking back down these lines of causation, our perceptions are coloured by the certain knowledge that they led to a bad outcome. As a consequence, these factors take on a sinister significance. But if we were to freeze any system in time without an accident having happened, we would see very similar imperfections, latent failures and technical problems. No system is ever perfect. The only thing that gives these same kinds of systemic weaknesses causal significance is that in a few

4

intensively-investigated events they were implicated in the accident sequence. If all that distinguishes these latent factors is the subsequent occurrence of a bad outcome, should we not limit our attention only to those proximal events that transformed such commonplace shortcomings into an accident sequence? In other words, should we not run with the legal and moral tide and simply concentrate on those individual failures having an immediate impact upon the integrity of the system? After all, no one is denying that 'sharp-enders' do make errors and commit procedural violations.

The remedial dimension

On the legal front, there is much that favours an individualistic approach. The scientific issues are unresolved. So where do we stand on the practical question of accident prevention? The answer here depends crucially upon two factors. First, whether or not latent organizational and managerial factors can be identified and corrected *before* an accident occurs, and, second, the degree to which these interventions can improve the system's natural resistance to local accident-producing factors.

One of the main aims of this chapter is to persuade readers that, for both of the above issues, the case for adopting a collectivist approach is extremely strong. However, this persuasion must be based on argument rather than hard evidence, since such collectivist measures have not been in place long enough to demonstrate proven gains in safety.

Introducing the theoretical framework

A recent survey of the Human Factors literature (Hollnagel, 1993) revealed that the estimated involvement of human error in the breakdown of hazardous technologies had increased fourfold between the 1960s and the 1990s, from minima of around 20% to maxima of more than 80%. During this period it has also become apparent that these contributory errors are not restricted to the 'sharp end', to the drivers, pilots, ships' officers, control room operators and others in direct control of a particular system. Nor can we only take account of those human failures that were the proximal causes of the accident. Major accident inquiries (e.g. Three Mile Island, *Challenger,* King's Cross, *Herald of Free Enterprise, Piper Alpha,* Clapham, *Exxon Valdez,* Kegworth, Dryden, etc.) indicate that the human causes of major accidents are distributed very widely, both within the organization as a whole and often over several years prior to the actual event.

The challenge facing those concerned with analysing the causes of such 'organizational accidents' is to develop a theoretical framework that can be meaningfully applied retrospectively to particular events such as the Dryden crash, as well as proactively to the whole gamut of complex, well-defended socio-technical systems. Such a model has to make sense of the specific and the general case, of past accidents as well as future ones.

More importantly, to have credibility it must lead to improved remediation and prevention. Unlike theories in the natural sciences that are judged by the amount of experimental interest they provoke, theories in the safety sciences can only be assessed by the harsher criterion of their practical utility. Do the applications of such a theory make technological systems more resistant to their natural hazards? In short, can it lead to improved 'safety health'?

Developing a generic framework for the origins of organizational accidents is a daunting task. The accidents listed above (as well as numerous other comparable disasters) involved widely differing technologies and a great variety of highly specific details relating to cause and effect.

The initial building blocks: the first step

The only way to proceeed is to ask: what do all of these complex, well-defended technologies have in common? The remainder of the chapter will deal with possible answers to this question in a stepwise fashion. The first step in the theory-building process is shown in Figure 1.1.

All complex technologies operating in hazardous conditions possess the following elements: organizational processes and their associated cultures, a variety of different workplaces involving a variety of local conditions, and defences, barriers and safeguards designed to protect people, assets and the environment from the adverse effects of the local hazards. Each of these elements is developed further below.

Organizational processes

Although all organizations exist within a broader economic, political and legislative setting, it is meaningful to confine the basic theoretical elements to those factors over which any one organization could reasonably be expected to exercise some measure of direct control.

6

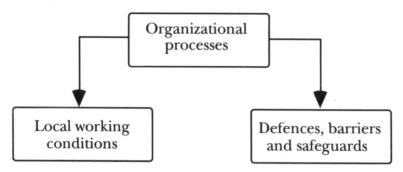

Figure 1.1. Showing elements common to all complex, well-defended technologies

Another reason for limiting the search for root causes to organizational processes is offered by Vaughan (1990), in her analysis of the regulatory failures contributing to the *Challenger* disaster:

> All organisations are, to varying degrees, self-bounded communities. Physical structure, reinforced by norms and laws protecting privacy, insulates them from other organisations in the environment. The nature of transactions further protects them from outsiders by releasing only selected bits of information in complex and difficult-to-monitor forms. Thus, although organisations engage in exchange with others, they retain elements of autonomy that mask organisational behaviour. (Vaughan, 1990)

Organizational culture is a widely used but variously-defined term. For our present purposes, the following definitions will serve to bracket our present use of the term:

> The culture of an organisation may be defined as the set of rarely articulated, largely unconscious beliefs, values, norms, and fundamental assumptions that the organisation makes about itself, the nature of people in general, and its environment. In effect, culture is the set of "unwritten rules" that govern "acceptable behaviour" within and outside the organisation. (Mitroff *et al.*, 1989)

> Shared values (what is important) and beliefs (how things work) that interact with an organisation's structures and control systems to produce behavioral norms (the way we do things around here). (Uttal, 1983)

Culture, then, comprises those attitudes and beliefs that both emerge

7

from and shape the way in which a company carries out its core business processes. These processes – all of which entail decision making, often at the highest levels – include the activities listed in Table 1.1.

Table 1.1.   Indicating some of the major processes common to all technological organizations

Goal-setting	Communicating
Policy-making	Designing/specifying
Organizing	Purchasing
Forecasting	Supporting
Planning	Researching
Scheduling	Marketing
Managing operations	Selling
Managing maintenance	Information-handling
Managing projects	Motivating
Managing safety	Monitoring
Managing change	Checking
Financing	Auditing
Budgeting	Inspecting
Allocating resources	Controlling, etc.

Cultural factors take a long time to develop and are slow to change. Their influence is widespread and pervasive. Cultural influences are disseminated throughout the organization in various ways, and colour the attitudes and behaviour of the workforce.

High-level decisions are shaped by external economic and political factors. However, for the purposes of this analysis, they represent the common starting point for the various failure pathways. They are the root causes, the basic causal *types* that subsequently give rise to specific *tokens* later in the accident development process (Wagenaar *et al.,* 1990).

It is taken as axiomatic that strategic decisions will carry some negative safety consequences for some part of the system. This is not to say that all such decisions are flawed, though some of them will be. But even those decisions judged at the time by reasonable criteria as good ones will carry a potential downside, particularly when they result in the unequal allocation of resources, or when they involve assumptions about uncertain futures.

It should be emphasized that this is not a question of shifting blame from flight crews and air traffic controllers to the boardroom. Rather,

8

it acknowledges that no matter how hard we try, we cannot hope to eliminate all the bad consequences of strategic decisions. The important thing is not so much to prevent these resident pathogens from being spawned in the first place, but to make their negative consequences *visible* to those who manage and operate the system.

Among other things, the strategic apex of the organization is responsible for (a) designing, equipping and managing the various places in which work is carried out, and (b) providing defences-in-depth against the foreseeable organizational hazards. We will consider each of these below.

Local working conditions

Local working conditions are the factors that influence the efficiency and reliability of human performance in a particular work context. The latter could be the flight deck of an aircraft, an air traffic control centre, a maintenance hangar engaged in operational repairs or major overhauls, a Boeing assembly plant in Seattle – any place, in fact, where groups of workers and their supervisors are engaged in the core business of the aviation industry, often (but not always) in close proximity to the local hazards.

The theory asserts that the negative consequences of top-level decisions (e.g. inadequate budgets, deficient planning, under-manning, commercial and operational time pressures, etc.) are transmitted along various departmental and organizational pathways to the different workplaces. There, they create the local conditions that promote the commission of unsafe acts. Many of these unsafe acts will be committed, but only very few of them will penentrate the defences to bring about damaging outcomes.

For the purposes of examining how local working conditions contribute to unsafe acts (errors and violations), it is convenient to divide them into two interacting clusters: those factors relating to the task and its immediate environment, and those relating to people's mental and physical states. These task-related and personal factors can each be further subdivided into three groups: error factors, violation factors and factors common to both errors and violations.

This categorization is shown in Figure 1.2, and some of the principal factors are listed below in Tables 1.2 and 1.3. The lists of factors in Tables 1.2 and 1.3 are not exhaustive, but they do indicate the major situational influences upon the commission of errors and violations.

The theory argues that all organizational failures are capable of creating the local conditions necessary for both errors and violations. It is only as we move down the causal sequence from organizational

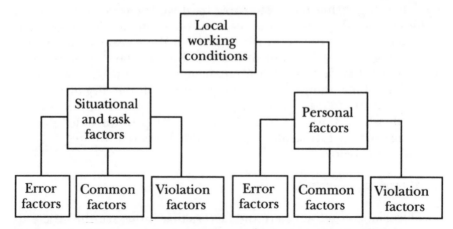

Figure 1.2. Showing the breakdown of local working conditions

processes to local working conditions and defences that these provocative factors become more easily discriminated, and even then there are a substantial number of situational and personal factors common to both errors and violations.

Defences, barriers and safeguards

Measures aimed at removing, mitigating or protecting against operational hazards now consume a very large part of the resources of organizations engaged in potentially dangerous activities. So diverse and widespread are these defensive measures that it is difficult to make a clear distinction between them and the non-defensive parts of a system.

On the face of it, the two most important productive elements in high-tech process plants (i.e. nuclear power and chemical plants) are the automated control systems and the human operators. Yet both perform essential defensive functions. The automation is there to increase the efficiency and safety of the plant by taking over functions previously left to variable and fallible human operators. The operators are there to restore the plant to a safe state in accident scenarios that go beyond the design basis of the smart technology. The same is true – to a large extent – of the flight decks of modern commercial aircraft (e.g. 747-400s and Airbus 340s).

Although current computer technology is very smart, it still cannot think 'on its feet' as well as human beings. So it is left to fallible people to cope with unexpected emergencies under less than ideal conditions; something they can do better than

Table 1.2.   Situational and task factors

Error factors	*Common factors*	*Violation factors*
Change of routine	Time shortage	Violations condoned
Negative transfer	Inadequate tools and	Compliance goes
Poor signal–noise ratio	equipment	unrewarded
Poor human–system	Poor procedures and	Procedures protect
interface	instructions (ambiguous	system not person
Poor feedback from	or inapplicable)	Little or no autonomy
system	Poor tasking	Macho culture
Designer–user mismatch	Inadequate training	Perceived licence to
Educational mismatch	Hazards not identified	bend rules
Hostile environment	Undermanning	Adversarial industrial
Domestic problems	Inadequate checking	climate (them and us)
Poor communications	Poor access to job	Low pay
Poor mix of hands-on	Poor housekeeping	Low status
work and written	Bad supervisior–worker	Unfair sanctions
instructions (i.e. too	ratio	Blame culture
much reliance on	Bad working conditions	Poor supervisory
knowledge in the head)	Inadequate mix of	example
Poor shift patterns and	experienced and	Tasks affording easy
overtime working	inexperienced workers	shortcuts

computers, but not always particularly well (see Reason, 1990). On some occasions, human controllers do come up with inspired solutions (i.e. Sioux City, Davies-Besse, etc.), but on others they can exacerbate an already dangerous situation (TMI, Chernobyl, Ginna, etc.). To compound the irony further, it is often the case that the operators are hampered in these less successful events by inadequacies in the engineered and procedural defences (Perrow, 1984; Roberts and Gargano, 1990).

Defences, barriers and safeguards can be classified along two relatively independent dimensions: (a) the functions served, and (b) modes of application within an organization.

Functions

- To create *awareness* and understanding of the risks and hazards.
- To *detect* and *warn* of the presence of off-normal conditions or imminent dangers.
- To *protect* people and the environment from injury and damage.
- To *recover* from off-normal conditions and to restore the system to a safe state.

11

Table 1.3. Personal factors

Error factors	Common factors	Violation factors
Attentional capture	Insufficient ability	Age and gender
Preoccupation	Inadeqate skill	High risk target
Distraction	Skill overcomes danger	Behavioural beliefs
Memory failures	Unfamiliarity with task	(gains outweigh risks)
Encoding interference	Age-related factors	Subjective norms
Storage loss	Poor judgement	condoning violations
Retrieval failure	Illusion of control	Perceived behavioural
Prospective memory	Least effort (cognitive	control
Strong motor programs	economics)	Personality
Frequency bias	Overconfidence	Non-compliant
Similarity bias	Performance anxiety	Unstable extravert
Perceptual set	(deadline pressures)	Low morale
False sensations	Arousal state	Bad mood
False perceptions	Monotony and boredom	Job dissatisfaction
Confirmation bias	Emotional states	Attitudes to system
Situational unawareness		Management
Incomplete knowledge		Supervisors
Inaccurate knowledge		Discipline
Inference and reasoning		Misperception of
Stress and fatigue		hazards
Disturbed sleep patterns		Low self-esteem
Error proneness		Learned helplessness

- To *contain* the accidental release of harmful energy or substances.
- To enable the potential victims to *escape* out-of-control hazards.

Modes of application

- Engineered safety devices (flight management systems, terrain warnings, automatic detection and shutdown, etc.).
- Policies, standards and controls (administrative and managerial measures designed to promote standardized and safe working practices.
- Procedures, instructions and supervision (measures aimed at providing local task-related know-how).
- Training, briefing, drills (the provision and consolidation of technical skills, safety awareness and safety knowledge).
- Personal protective equipment (anything from safety boots to space suits).

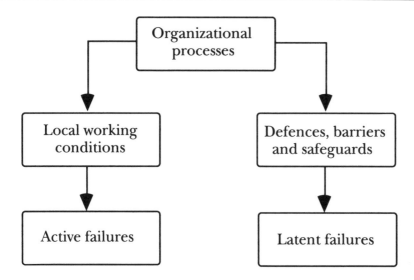

Figure 1.3. Active and latent failures: the next step in the theory-building process

There is no such thing as a perfect set of defences against all eventualities. Many of the shortcomings will be the direct consequence of organizational decision processes. Such imperfections would include the absence of necessary defences and weaknesses in existing defences. Together they comprise the bulk of the system's latent failures.

Active and latent failures: the second step

When people are involved in a complex system, there will be failures. These may occur in either the workplace or in relation to the defences. Figure 1.3 describes the next step in the theory-building process. It shows active failures occurring in the various workplaces. Latent failures, on the other hand, are mainly associated with weaknesses in or absences of defences.

Active and latent failures are distinguished in two ways. The first is by the length of time it takes for these failures to reveal their adverse effects upon the integrity of the system. Active failures have an immediate and direct impact, whereas latent failures may lie dormant for long periods, sometimes many years, before they combine with active failures and local triggering events to breach the system's defences.

13

The second distinction, clearly dependent on the first, relates to who creates these failures. Active failures are committed by those in direct contact with the system: pilots, air traffic controllers, maintenance mechanics and the like. Latent failures, however, derive from decisions taken in the managerial and organizational spheres. These are people separated in both time and space from the immediate human–system interfaces.

Human active failures are errors or violations committed by those at the sharp end of the system. Usually, the consequences of these active failures are caught by the system defences or by the perpetrators themselves, and have no ill effects. On some occasions, they may occur in conjunction with a breach in the defences and cause an accident. The less defended the system, of course, the more likely it is that active failures will have immediate bad outcomes. On other occasions, active failures may themselves create instant gaps in the defences, as at Dryden where the flight crew chose not to have the aircraft de-iced before take-off.

Latent failures are loopholes in the system's defences, barriers and safeguards whose potential existed for some time prior to the onset of the accident sequence, though usually without any obvious bad effect. On some occasions, however, these weaknesses may combine with both active failures or local triggers (or frequently both) to create a trajectory of accident opportunity (sometimes only momentary) through some or all of the system's various protective layers. It is these temporarily lined-up gaps in the various defences-in-depth that constitute an event.

Most latent failures are only discovered once a defence or barrier has failed. However, this does not mean that it is necessary to wait until after a full-blown accident before they can be made manifest. The normal run of operations fortunately offers a fairly large number of 'free lessons' in which a defence or barrier is shown to be deficient without adverse consequences.

While many latent failures are only revealed retrospectively, the *potential* for a system to develop latent failures may be assessed proactively. The number of latent failures in a system will be a function of the overall 'safety health' of the organization. Safety health can be inferred at regular intervals by making assessments of a limited number of key parameters of organizational activity. This will be discussed at a later point in this book.

Active and latent failures also differ in the necessary basis for their classification. Whereas active failures are categorized according to their psychological origins, latent failures must be described in systemic terms. These two classifications are explained below.

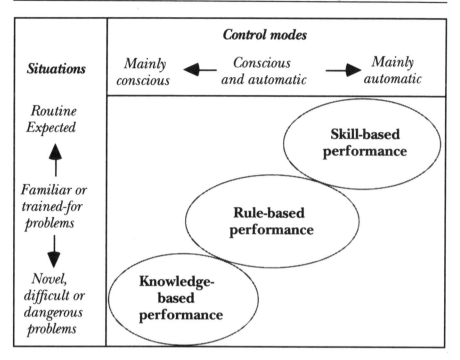

Figure 1.4. Locating the performance levels in the activity space

Classifying active failures

To make a workable classification of human failures, it is first necessary to distinguish the different levels at which people perform their actions. There are three such levels: *skill-based, rule-based* and *knowledge-based.*

Human behaviour is shaped by both personal (or psychological) and situational factors. Thus, we can distinguish the three performance levels along two dimensions, one having to do with the way we control our actions, and the other relating to whether or not the situation in which they occur is routine or problematic.

These two dimensions – *conscious-automatic* and *routine-problematic* – define an 'activity space' onto which we can map the three performance levels: skill-based, rule-based and knowledge-based. This is shown in Figure 1.4.

- At the *skill-based level,* we carry out routine, highly-practised tasks in a largely automatic fashion, except for occasional conscious checks on progress.

15

- We switch to the *rule-based level* when we notice a need to modify our largely preprogrammed behaviour. We have to take account of some change in the situation. This problem is often one that we have encountered before and for which we have some prepackaged solution (i.e., one developed by training and/or experience). It is called the rule-based level because we apply stored rules of the kind: *If (this situation) then do (these actions)*. In applying these stored solutions we operate very largely by unconscious pattern-matching: we automatically match the signs and symptoms of the problem to some stored solution. We may then use conscious thinking to check whether or not this solution is appropriate.

- The *knowledge-based level* is something we come to very reluctantly. Only when we have repeatedly failed to find a solution using known methods do we resort to the slow, effortful and highly error-prone business of thinking things through on the spot. Given time and the freedom to explore the situation with trial and error learning, we can often come up with good solutions. But people are not usually at their best in an emergency – though there are some notable exceptions. Quite often, our knowledge of the problem situation is patchy, inaccurate, or both. Consciousness is very limited in its capacity to hold information, usually not more than two or three distinct items at a time. It also behaves like a leaky sieve, forgetting things as we turn our attention from one aspect of the problem to another. In addition, we can be plain scared, and fear (like other strong emotions) has a way of replacing reasoned action with 'knee jerk' or over-learned reactions.

The three performance levels provide a principled basis for classifying both errors and procedural violations. Before presenting this classification, however, it is worth outlining the main differences between errors and violations.

Errors may be defined as the failure of planned actions to achieve their desired consequences. This failure may occur in one of two ways. The plan is adequate, but the actions deviate from the plan (slips, lapses, fumbles); or the actions may conform to the plan, but the plan is not appropriate for achieving its desired ends (mistakes). Thus the first type of error is a failure of execution, and the second is a failure at the level of formulating the intention or plan.

Violations are deviations from safe operating practices, procedures, standards or rules. Such deviations can either be deliberate or erroneous (i.e. speeding without being aware of either the speed or

Table 1.4. Relating error types and violation types to performance levels

Performance levels	*Error types*	*Violation types*
Skill-based	Slips and lapses	Routine violations
Rule-based	RB mistakes	Situational violations
Knowledge-based	KB mistakes	Exceptional violations

the restriction). However, we are mostly interested in deliberate violations, where the actions (though not necessarily any bad consequences) were intended.

Violations differ from errors in a number of important ways:

- Errors are unintended. Violations are deliberate. The deviation from procedures is intended, though the occasional bad consequences are not.
- Whereas errors arise primarily from *informational problems* (i.e. forgetting, inattention, incomplete knowledge, etc.), violations are more generally associated with *motivational problems* (i.e. low morale, poor supervisory example, perceived lack of concern, the failure to reward compliance and sanction non-compliance, etc.).
- Errors can be explained by what goes on in the mind of an *individual*, but violations occur in a regulated *social* context.
- Errors can be reduced by improving the quality and the delivery of necessary information within the workplace. Violations require motivational and organizational remedies.

Both errors and violations can be classified into three distinct types, according to the level of performance at which they occur. These error and violation categories are shown in Table 1.4.

Skill-based slips and lapses   These are actions-not-as-planned that occur during the execution of well-practised and familiar tasks, in which our movements are largely automatic with only intermittent checks on progress by conscious attention. Slips and lapses are associated with different kinds of executive failure.

- *Attentional slips* in which we fail to monitor the progress of our routine actions at some critical choice point, often following a change in either our intentions or the surrounding

17

circumstances. The upshot is that we do what is customary or habitual in those circumstances rather than what was then intended. In other words, we make 'strong-but-wrong' errors.

- *Memory lapses* in which we omit items in the plan of action, or forget what it is we earlier intended to do. Memory lapses of one kind or another constitute the commonest kind of everyday absent-minded error, and are of considerable significance in aircraft maintenance where omissions during reassembly represent the largest category of active failures (see Reason, 1994).

- *Perceptual errors* in which we misrecognize some object or situation. Here, expectation and habit can play a large part. Many train accidents, for example, have been due to the driver expecting (on the basis of past experience) to see a green signal at that point on the track, whereas its actual aspect was red. In aviation, misreadings of the old 3-pointer altimeter – the so-called killer altimeter – have been implicated in many serious crashes (Reason and Mycielska, 1982). Also included in this category are the various forms of disorientation that arise because our land-based position and motion senses give false signals due to the unusual force environment encountered in 3-dimensional flight.

Slips and lapses are the penalties we pay for being able to automatize our actions and perceptions. Normally, they have trivial consequences – as when we get into the bath with our socks on, or say 'thank you' to a stamp machine, or try to open a friend's front door with our own latch key. But the same errors can have markedly different consequences according to their circumstances of occurrence. In the kitchen, switching on the kettle instead of the coffee machine is merely embarrassing. On the flight deck, hitting the wrong switch or shutting down the wrong engine can be catastrophic. The basic errors are the same, but situations vary considerably in the degree to which these similar slips and lapses are likely to be forgiven.

Rule-based mistakes Rule-based (RB) mistakes involve the application of prepackaged but inappropriate solutions to problems that people have either encountered many times before or which they have been trained to handle. A problem here is anything that requires some alteration of the current routine behaviour.

RB mistakes divide into two broad categories: the *misapplication of normally good rules*, and the *application of bad rules*. We will consider each of these below.

18

- *Misapplication of good rules.* Here, a 'good rule' is one that has proven utility in a particular situation. These useful problem-solving rules become an established part of a person's expertise. However, occasions arise in which the rule is wrongly applied. This usually happens in situations that share many common features with the usual problem situation, but in which there are counter-indications that are overlooked. An everyday example is braking to avoid a pedestrian stepping out into the road, but overlooking the fact that the road surface is icy. Here we have the application of a 'strong-but-wrong' rule.
- *The application of bad rules.* In the course of learning a job, most people pick up some 'bad' problem-solving rules. They are 'bad' because they are inelegant, inadvisable or can lead to something going wrong later. However, these 'bad rules' achieve their purpose a lot of the time, so they become established as a part of the person's problem-solving repertoire.

These two types of RB mistakes have quite different origins. The misapplication of good rules arises from a failure to discriminate between appropriate and inappropriate problem situations. It must be appreciated that these differences can be quite small and hard to spot. There are often many features in common. Also, we tend to apply solutions to familiar problems on the basis of largely automatic pattern-matching. A problem diagnosis will spring to mind and only then we may think about how appropriate it is.

The bad rules become a part of a person's toolbag of solutions because circumstances are often relatively forgiving. Less-than-ideal ways of tackling problems can bring about workable though undesirable solutions. The technician carrying out the rewiring of the Clapham signal box was found to employ a number of bad practices because he had neither been properly trained nor closely supervised.

Knowledge-based mistakes KB mistakes occur when a person is attempting to solve a novel problem (i.e. one for which his or her training and/or experience has not provided a pre-programmed solution). This entails conscious on-line reasoning rendered error-prone by the limited capacity of working memory and the use of incomplete or inaccurate mental models of the problem situation. Such 'thinking on one's feet' is also prey to confirmation bias (jumping to the wrong conclusion and then bending the facts to fit the conclusions), over-confidence, similarity bias (crudely matching like to like) and frequency bias (a proneness to call to mind frequently encountered scenarios or possibilities).

19

Violations at the skill-based level   Violations at the skill-based level form part of a person's repertoire of skilled actions. They often involve corner-cutting (i.e. following the path of least effort between two task-related points). Such routine violations are promoted by a relatively indifferent environment. That is, one that rarely punishes violations or rewards compliance with the rules.

Violations at the rule-based level   Safety procedures, rules and regulations are written primarily to control behaviour in problematic or risky situations, and are thus most abundant at the RB level. In the initial stages of a particular system or technology, the procedures may simply provide instructions on how to carry out the necessary tasks and how to deal with foreseeable hazards (see Figure 1.5). But procedures are continuously being amended to incorporate the lessons learned in past incidents and accidents. Such modifications usually proscribe particular actions known to have been implicated in a particular accident or incident scenario.

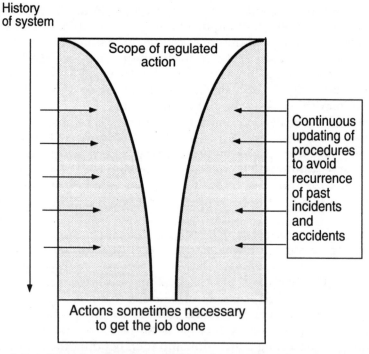

Figure 1.5.   **Showing the diminishing scope of action as procedures are modified to minimize the recurrence of past incidents and accidents**

20

The upshot is that the scope of allowable action gradually diminishes as the system or technology matures. However, the range of actions necessary to get jobs done within current operational and commercial constraints may not diminish. This creates the conditions for necessary or situational violations. These are problems for which violations offer possible or, in some cases (e.g. Chernobyl), the only solutions.

RB violations are likely to be more deliberate than SB violations. However, just as mistakes are intentional, actions carried out in the belief that they will achieve their desired ends, so situational violations tend to be deliberate acts carried out in the belief that they will not result in bad consequences. These violations are shaped by cost-benefit trade-offs, where the benefits are seen as outweighing the possible costs.

Such assessments can be mistaken. Thus, situational violations can involve both mistakes and procedural deviations, a class of violation that we might call 'misventions' (a blend of mistaken circumventions).

Violations at the knowledge-based level By definition, activities at the KB level take place in atypical or novel circumstances for which there is unlikely to be any specific training or procedural guidance. Trainers and procedure writers can only address known or foreseeable situations.

The Chernobyl disaster (Reason, 1987) provides perhaps the best documented account of exceptional violations. The experimental plan already demanded a situational violation, namely operating without the emergency core cooling system for the purpose of testing the voltage generators. The sequence of active human failures began with a slip (the undershooting of the 25% power level) and then proceeded on the basis of a misvention (continuing the tests while the RBMK reactor was operating at considerably less than the minimum 20% power level, thus making it liable to positive void coefficient). Thereafter followed a series of exceptional violations by which the operators successively shut off safety systems, making the explosions inevitable. They did so in apparent ignorance of the basic physics of the plant, and in the hope of completing the tests in an already diminished window of opportunity.

Problems encountered at the KB level do not have to be novel in the sense that the surface of Mars would be novel to some future astronaut. Quite often they involve the unexpected occurrence of a rare but trained-for situation, or an unlikely combination of individually familiar circumstances.

21

Mode

Function	Engineered safety features	Standards, policies, controls	Procedures, instructions supervision	Training, briefings, drills	Personal protective equipment
Awareness					
Detection warning					
Protection					
Recovery					
Containment					
Escape					

Figure 1.6.   A matrix for locating specific latent failures

Classifying latent failures

Latent failures may be categorized by the function and mode dimensions of defences, barriers and safeguards, described above. Together, they create a matrix within which it is possible to locate specific latent failures. This matrix is shown in Figure 1.6.

The matrix provides a means not only for classifying latent failures in general, but also for mapping the relative densities of latent failures in specific accidents. Each accident has its own characteristic 'thumbprint' of latent failures.

By the same token, it is also possible to use the latent failure matrix as a systematic basis for evaluating the effectiveness of a system's safety management system. A cell by cell analysis would allow the identification of areas of potential vulnerability.

The matrix is confined to a particular organization and thus makes no mention of the crucial defensive role played by external regulators. Failures of both self-regulation and external regulation played a crucial part in the *Challenger, Herald* and *Piper Alpha* accidents (Vaughan, 1990; Sheen, 1987; Cullen, 1990). However, regulation serves the same defensive functions as those listed for the intra-organizational modes, and could thus be represented as a separate column external to the matrix.

Accidental events: the third step

The next step in developing the accident causation model concerns the event itself. An event is defined here as a complete or partial penetration of an accident trajectory through the system's defensive layers. An accident trajectory is something that, if it penetrated all the defences, barriers and safeguards, could bring uncontrolled hazards in contact with potential victims. The third step of the model development is shown in Figure 1.7.

It is at the event level that the active and latent failure pathways come together to create complete or partial trajectories of accident opportunity. The causal pathways can also interact with local triggers. These include such factors as atypical system states (i.e. startups or shutdowns), local environmental conditions (i.e. the spring tides at Zeebrugge) or adverse weather (i.e. the low temperatures on the night preceding the *Challenger* launch).

Figure 1.8 shows the nature of an event in more detail. The gaps in the defences are of various kinds:

- Longstanding gaps due to dormant weaknesses or undiscovered shortcomings in the defences. They are revealed as latent failures by events of varying consequences.
- Gaps created knowingly during the course of maintenance or atypical system states.
- Gaps created by active failures, either unsafe acts or the breakdown of particular components. The deliberate disabling of an engineered safety feature or the violation of safe operating procedures represent examples of these recently created loopholes.

In activities possessing minimal defences (e.g. cutting a seismic line through jungle), a single error (misjudging the swing of the machete) may be sufficient to bring about adverse consequences (a badly cut leg, for example). In most modern, well-defended systems, however, a complete accident trajectory requires the precise lining up of holes in all the defensive layers. The probability of such correspondences is

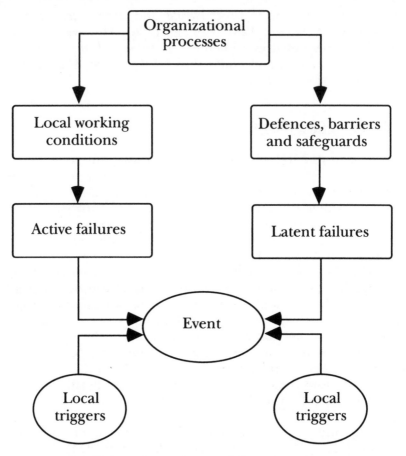

Figure 1.7.   Showing active and latent failures combining to cause an event

very small, though human beings (and particularly maintenance technicians – see Rasmussen, 1980) have a knack for creating common mode failures and hence a colinearity of gaps.

The event level is highly dynamic. Gaps appear, disappear and reappear. They shrink and expand. They change their locations within the defensive layer. As elsewhere in the model, stochastic influences are continually at work.

This shifting picture also includes large numbers of unsafe acts whose adverse consequences are detected and checked by the relevant defences or by the perpetrators themselves. Only relatively few of them will find chinks in the defences in well-protected systems. Potential accident trajectories may penetrate one or two of the defensive layers

24

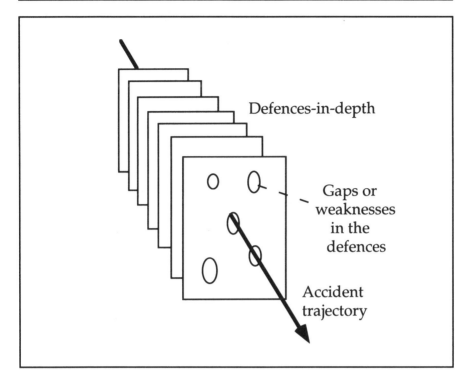

Figure 1.8. An event involving the complete penetration of the system's defences, barriers and safeguards

and then vanish, often unnoticed and unrecorded.

It should also be appreciated that, although Figure 1.8 shows layers of successive barriers, not all systems possess such defences-in-depth, nor is it always necessary for an accident trajectory to penetrate all the layers to achieve its damaging consequences.

The consequences of an event can vary from the catastrophic to the 'free lesson'. All, however, provide crucial learning experiences for the system in question, since each event reveals hitherto unnoticed or uncorrected latent failures. A prerequisite for appropriate learning is a safety information system that identifies not just the proximal active failures, but also the latent failures and their parent organizational pathogens.

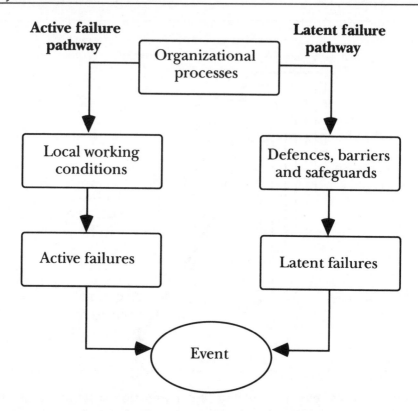

Figure 1.9. Showing active and latent failure pathways

Causal pathways: the fourth step

An important feature of the causal model is a clear separation between the active and latent failure pathways, though both have a common origin in the strategic organizational processes. The two pathways are shown in Figure 1.9.

The reason for distinguishing these pathways is to reflect the differing character of past and potential accidents. In some cases – *Torrey Canyon, Exxon Valdez,* Kegworth, the BAC1-11 windscreen incident, the Purley train crash, etc. – the accident trajectory ran mainly along the active failure pathway. Other accidents, such as *Challenger, Piper Alpha* and King's Cross, involved little or no active failures as proximal causes. In these cases, the accident trajectory passed through longstanding latent failures in the defences. A large proportion of accidents, however, require the timely concatenation of both active and latent failures to achieve a complete trajectory of accident opportunity. This was true of Flixborough, the Tenerife runway collision, Three Mile Island, Ginna, Chernobyl, Bhopal, Heysel, Clapham, the *Herald of Free Enterprise*, the Philips 66 fire in Texas, and the Dryden crash.

Are some types of system more prone to active failure trajectories, and others more prone to latent failure trajectories? Although the listed examples above indicate that no hard and fast line can be drawn, it is tempting to suggest that accidents to transport systems (ships, aircraft, trains, cars, etc.) nearly always require some contribution from the active failure pathway. The most commonly cited cause of commercial aviation accidents is controlled flight into terrain (CFIT) discussed later in the book. The vulnerability of transport systems to active failures stems largely from their speed and the proximity of other hazards, many of which are equally mobile. They tend to have a very short time scale of accident development.

By contrast, process plants, particularly nuclear power plants, can withstand relatively long intervals between the onset of an event and the likelihood of some damaging outcome. This tolerance arises partly from the dynamics of the process and partly from a combination of designed-in defences, such as the diversity and redundancy of safety-critical components, as well as successive layers of barriers and safeguards. However, it is just these features, with their associated complexity and opacity, that render modern process plants especially prone to the insidious accumulation of organizational pathogens and latent failures. Such well-defended installations are largely proof against single active failures, either human or mechanical. Large numbers of coincident latent failures, either with or without the co-occurrence of active failures, are usually necessary to achieve a 'successful accident' in systems of this kind.

Defences stand between hazards and damaging losses (of life, health, property, money, the quality of the environment, etc.). In some instances, these defences are undermined from within the system via the active failure pathway. In other cases, the gradual accretion of latent failures reaches a point where the hazards can invade the system from 'outside' (although not always in the literal sense). In Figure 1.8, therefore, one can draw the direction of the accident trajectory as running either from inside out (the active failure pathway) or from the outside in (the latent failure pathway), depending upon whether active or latent failures were the principal proximal factors.

27

Summarizing the theory

Organizational processes – decisions taken in the higher echelons of the system – seed *organizational pathogens* into the system at large. These resident pathogens take many forms: managerial oversights, ill-defined policies, lack of foresight or awareness of risks, inadequate budgets, lack of legal control over contractors, poor design, specification and construction, deficient maintenance management, excessive cost-cutting, poor training and selection of personnel, blurred responsibilities, unsuitable tools and equipment, commercial pressures, missing or flawed defences and the like. The adverse consequences of these pathogens are transported along two principal pathways to the various workplaces, where they act upon the *defences, barriers and safeguards* to create *latent failures* and upon *local working conditions* to promote *active failures* (errors, violations and component failures).

Subsequently, these active and latent failures may act to create an *event* (a complete or partial trajectory through the defensive layers). Events may arise from a complex interaction between active and latent failures, or from factors present predominantly in one or other pathway. Both local triggering factors and random variations can assist in creating trajectories of accident opportunity.

The remedial implications of this theoretical framework have been pointed out at various points throughout this chapter. The remedial applications of the theory are both proactive and reactive. By specifying the organizational and situational factors involved in the causal pathways, it is possible to identify potentially dangerous latent failures *before* they combine to cause an accident. The same framework can also be used in reverse to track back from some incident or accident, via the active and latent failure pathways to their organizational roots. Hitherto, this has only been achieved through long and expensive legal inquiries. Now the conceptual tools exist to do this for any event, no matter how trivial.

References

Cullen, The Hon. Lord (1990) *The Public Inquiry into the Piper Alpha Disaster* 2 vols. London: HMSO.

Hidden, A. (1989) *Investigation into the Clapham Junction Railway Accident.* London: HMSO.

Hollnagel, E. (1993) *Human Reliability Analysis; Context and Control.* London: Academic Press.

Mitroff, I. I., Pauchant, T., Finney, M., and Pearson, C. (1990) Do some organizations cause their own crises? The cultural profiles of crisis-prone vs. crisis-prepared organizations. *Industrial Crisis Quarterly,* **3**: 269–283.

Moshansky, The Hon. V.P. (1992) *Commission of Inquiry into the Air Ontario Crash at Dryden, Ontario.* Ottawa: Ministry of Supply and Services, Canada.

Perrow, C. (1984) *Normal Accidents: Living with High-Risk Technologies.* New York: Basic Books.

Rasmussen, J. (1980) What can be learned from human error reports? In: K. Duncan, M. Gruneberg and D. Wallis (Eds), *Changes in Working Life.* London: Wiley.

Reason, J. (1987) The Chernobyl errors. *Bulletin of the British Psychological Society,* **40**: 201–206.

Reason, J. (1990) *Human Error.* New York: Cambridge University Press.

Reason, J. (1994) Comprehensive error management in aircraft engineering. In *Human Factors in Maintenance. Proceedings of the Aerotech 94 Congress.* London: Institute of Mechanical Engineers.

Reason, J. and Mycielska, K. (1982) *Absent-Minded? The Psychology of Mental Lapses and Everyday Errors.* Englewood Cliffs, NJ: Prentice-Hall.

Roberts, K. H. and Gargano, G. (1990) Managing a high-reliability organisation: A case for interdependence. In: M. von Glinow and S. Mohrman (Eds), *Managing Complexity in High Technology Organizations.* Oxford: Oxford University Press.

Sheen, Mr. Justice (1987) *MV Herald of Free Enterprise. Report of Court No. 8074.* London: Department of Transport.

Uttal, B. (1983) The corporate culture vultures. *Fortune,* October 17.

Vaughan, D. (1990) Autonomy, interdependence, and social control: NASA and the Space Shuttle *Challenger. Administrative Science Quarterly,* **35**: 225–257.

Wagenaar, W.A., Hudson, P.T.W., and Reason, J. (1990) Cognitive failures and accidents. *Applied Cognitive Psychology,* **4**: 273–294.

29

2 Erebus and beyond

Old and new perspectives

Theory and practice

This chapter reviews an accident to a DC-10 aircraft which occurred in November 1979, at Mount Erebus in the Antarctic. Before addressing the accident itself, a number of topics are briefly considered, with a view to helping the reader evaluate how best to bridge the gap between abstract and applied approaches to systemic safety. The Erebus accident is subsequently used to illustrate the analytic benefits of the principle and ideas outlined in Chapter 1. The chapter concludes with a brief re-examination of conceptual issues which arise from the earlier discussion.

Organizational risk management

Modern organizations face a formidable range of risks, ranging from strategic changes in the commercial environment, through to the adverse commercial impact of accidents and public relations disasters. It is a truism that effective organizations actively attempt to manage those risks which potentially impact upon organizational survivability. The difficult task for management is to determine which risks carry the most potent dangers (Fischhoff, 1994; Hood et al., 1992). In addition, management must also establish suitable structures and methods for anticipating, containing and controlling the impact of those unexpected and unanticipated events which potentially threaten their organization.

Human error and systemic risk management

It is some time since Cicero uttered the remark that 'to err is human'. Human error has been with us for as long as humans have inhabited the planet, and it certainly seems destined to remain with us for some time to come. Human error is frequently a precursor to failures of risk management systems, whether these are precipitated by omissions, or by unsafe acts. From a risk management viewpoint, an inability to absorb the consequences of unsafe acts and omissions is ultimately considered to be a symptomatic failure of the overall risk management system (Grose, 1987). From this perspective, an accident can ultimately be deemed to derive from a system which is inadequately specified or designed, or which has insufficient 'defences-in-depth' (Chapter 1; International Civil Aviation Organisation, 1994).

Managerial imperatives and realities

Words are cheap, and reiterating truisms and a received wisdom dating back to Cicero serves only to again repeat sentiments which have been regularly voiced over the years. Managers are practical people, with limited time and resources. Urging them to 'manage risk effectively' is somewhat analogous to telling pilots that they should avoid 'pilot error'. Most of us, whether manager or front-line employee, have good intentions. Actually achieving our goals and good intentions in practice is somewhat more problematic.

Real world organizational situations and contingencies seem impervious to easy risk management solutions. Actions based upon a simple and isolated desire to avoid human error, or aspirations to establish 'suitable risk management structures', frequently seem doomed to failure. For what is to be the substantive meaning and practical significance of such desirable intentions? Furthermore, human error is stubborn and enduring, so it is often the case that highly sophisticated discrete solutions to human deficiencies are promptly succeeded by even more highly sophisticated manifestations of human error.

Direct and indirect causes

When we look at the precipitating events which lead directly to accidents or unfortunate outcomes, there is a natural tendency to see the most proximate or immediate factors in the causal sequence as

31

being the direct 'cause' of the undesired outcome. The idea of a linear causal chain, with uninterrupted cause:effect relationships, sits easily with those from technical backgrounds.

When technical components fail, we rarely do more than to record the failure and perhaps consider how we might best improve our design, operating, maintenance or quality control procedures. However, when human performance fails in some way to meet expectations, we tend to judge that performance against an anticipated, assumed or imputed standard of behaviour. We often apportion responsibility and blame for the outcome (Alicke, 1992). We tend to make judgments of a qualitatively different kind than when considering failures of inert technical components – in other words, we tend to retrospectively judge human participants as autonomous and volitional individuals, and not as impersonal, inert, systemic components.

Accidents and the wisdom of hindsight

One consequence of this view is a tendency for investigators to treat existing social and organizational activities as irrelevant *standing conditions* (Einhorn and Hogarth 1986, Mackie 1974), and to reason backwards in a linear progression from the circumstances of an accident or incident. Frequently this retrospective reasoning is performed in the light of those rules and regulations deemed, *ex poste*, to have been applicable. Such reasoning inevitably finds a stage at which the accident causal sequence could have been broken. This invariably is a point at which an individual failed to act in accordance with precedent, rule or regulation. Apportioning responsibility is then reasonably straightforward and, in retrospect, the ensuing findings often appear blindingly 'obvious'.

Such an approach to human failure is common, and in appropriate circumstances it can be valuable and helpful. On the other hand, in different circumstances this approach can equally amount to little more than a superficial restatement of the relevant events, ending with some residual and trivial category of human failure. The historical epithet 'pilot error' comes to mind in this context. Few readers will need reminding that such retrospective explanatory assignations have rarely contributed much to the prevention of subsequent accidents.

Individual and organizational responsibilities

Important issues regarding *individual* and *organizational* responsibilities arise in this context. When we allocate specific responsibilities to workers, is it not their duty to perform in accordance with those responsibilities? Lord Denning (1978, p451B) appears to endorse this point of view in his often quoted judgment in a case involving the dismissal of a pilot:

> There are activities in which the degree of professional skill which must be required is so high, and the potential consequences of the smallest departure of that high standard are so serious, that one failure to perform in accordance with those standards is enough to justify dismissal.

At one extreme this can be viewed as equating to the notion of *absolute responsibility* – with consequences which the harsh findings of various 19th-century Courts of Enquiry into shipping accidents readily testify (Barnaby, 1968). Such views set a high standard for professional performance. On the other hand, recent years have seen a host of inquiries which implicate official bureaucracies and organizational functioning as key accident precursors. By way of comparison, consider the following extracts from two recent accident investigations:

> The reality was that the concern for safety was allowed to co-exist with working practices which were positively dangerous. This unhappy co-existence was never detected by management, and so bad practices were never eradicated. The errors go much wider and higher in the organisation than merely to remain at the hands of those who were working on that day. (Hidden, 1989)

> In this accident ...established company procedures were not being followed by personnel in the hangar. Inspectors, who were responsible for assuring the quality of work in accordance with established procedures, were among the worst offenders. ...the lax attitude of personnel in the hangar suggests that management did not establish an effective safety orientation for its employees. In fact, the failure of management to ensure compliance with air carrier policy must be considered a factor in the cause of the accident. (NTSB, 1992, p 44)

Analogous findings are also to be found in the Moshansky Inquiry into the Dryden F-28 accident (Moshansky, 1992; see also Chapter 3) and the *Herald of Free Enterprise* Inquiry (Sheen, 1987). Here again individual and organizational contributions to the accidents were identified. In fact, these are among a host of recent inquiries which have had to resolve a tension between manifest organizational deficiencies and the formal responsibilities of individual workers. Most of these inquiries have also found it necessary to address those long-term standing conditions – such as organizational culture, working practices, customs and attitudes – which combined with isolated unsafe acts by individuals to cause an accident (Anderson and Wreathall, 1990; Maurino, 1992).

Why do we investigate accidents?

Cicero supplemented his observation that 'to err is human' with the further observation that 'only a fool perseveres in error'. For as long as human imperfectability remains a fact of life, learning the relevant preventive lessons will be one of the primary reasons why aviation accidents are formally investigated, and in such painstaking detail. However, it is equally the case that there are sometimes confused and confounded objectives when incidents and industrial accidents are investigated. Investigations often serve purposes other than accident prevention – including organizational, political and legal purposes.

Management, for instance, may feel that they have to formally demonstrate that they are willing to move to prevent a recurrence of similar events, or that they must formally apportion blame (Johnston, in press). There may be a need to demonstrate that the public will be rendered safe from further occurrences of similar errors. In circumstances where death or significant injury is involved, relatives often seek a means by which their grief can be publicly atoned. They may also seek damages and, perhaps, retribution. Inquiries directed to these various purposes are rarely effective vehicles for accident prevention.

While in one sense obvious, it must nevertheless be clearly understood that clarity of purpose during an accident investigation is essential. It will also be obvious that how one looks at the events involved in an accident is to some extent influenced by one's values, investigative purposes, intentions and, especially, the chosen analytic model. The significance of these remarks will be endorsed by the issues arising in the following case study. It is thus axiomatic that the model offered in Chapter 1 is but one way of approaching accident analysis. It is the reader who must ultimately decide if the use below and in other chapters of this book of that model is coherent and persuasive. The reader must also decide if the discussion successfully demonstrates that the approach outlined in Chapter 1 offers tangible practical benefits in accident investigation and, preferably, prevention.

Erebus

Why Erebus?

The accident at Mount Erebus was chosen for review in this chapter for a variety of reasons. These include the fact that there were two authoritative investigations into the accident, each of which reached radically different conclusions. The findings of one inquiry reflect the then (1979) orthodoxy in accident investigation, whilst the findings of the second anticipated the growing investigative trend, as formalized in Chapter 1. An additional advantage of choosing this accident is that there is a substantial amount of supporting material and evidence available for review (a detailed bibliography is provided at the end of the chapter).

Another interesting and valuable feature of the Erebus accident is that neither investigation can prove the other wrong. This assertion may seem surprising, since we are conditioned to expect definitive answers from formal inquiries. However, one of the early lessons one learns from studying the evidence in this case is that there are at least two ways of looking at the events leading to this accident. And there are at least two ways of marshalling the evidence. So the salience of the 'facts' surrounding the circumstances of the accident turn out to be dependent upon the chosen analytic model and one's interpretation of the relevant events (Johnston, 1989). The different investigations of this accident thus provide a classic study in how values and interests shape the process of inquiry (Trusted, 1981; Trigg, 1985; Johnston, 1985).

A cautionary note

It will not be possible to do justice to the intricacies, nuances and complexity of this accident in the space available here. The reader should be alert to the fact that there are strongly held points of view on both sides of the debate as to the primary cause of this accident (Macfarlane, 1991). While the second investigation – a Royal Commission of Inquiry – takes formal precedence, there is a minority who feel strongly that the original accident investigation is correct.

It is also important to observe that no criticism of either investigation is intended in what follows. Nor is it implied that the airline or any of the other organizations involved in the accident, or in its investigation, were notably different from the norm in international aviation. Indeed, the general thrust of the overall analysis advanced here is that even the best run organizations are not immune to the lethal dangers of cumulative latent failures. Undoubtedly, deadly latent pathogens currently lie dormant in most organizations. The objective must be to learn how we might identify and eradicate these latent pathogens. In the light of this objective, the following review emphasizes how system safeguards failed on one occasion, and itself benefits from more than ten years of progress in safety thinking since the accident.

Erebus – the accident

The flight in question was one of a series of round trips from New Zealand to Antarctica for the purposes of sight-seeing (see Figure 2.1). No landing in Antarctica was ever contemplated. The facts surrounding the accident are deceptively simple: the flight crew descended to low altitude in an unfamiliar environment, quite different to that in which the airline conducted its regular operations and, at 1500 feet, collided with the ground while flying directly towards the base of Mount Erebus (elevation 13,500 feet). The aircraft was completely destroyed by the impact and subsequent fire. All occupants perished on impact.

Erebus – one accident, two inquiries

The formal accident report (*Aircraft Accident Report 79-139*) determined that the captain and crew were to blame for the accident, concluding:

> 3.37 Probable cause: The probable cause of this accident was the decision of the captain to continue the flight at low level toward an area of poor surface and horizon definition when the crew was not certain of their position and the subsequent inability to detect the rising terrain which intercepted the aircraft's flight path.

However a subsequent Commission of Inquiry (Report of the Royal Commission, 1981) concluded that collective organizational failures predominated:

> 393. In my opinion therefore, the single dominant and effective cause of the disaster was the mistake by those airline officials who programmed the aircraft to fly directly at Mt. Erebus and omitted to tell the aircrew.

Mr. Justice Mahon, the Royal Commissioner, subsequently added:

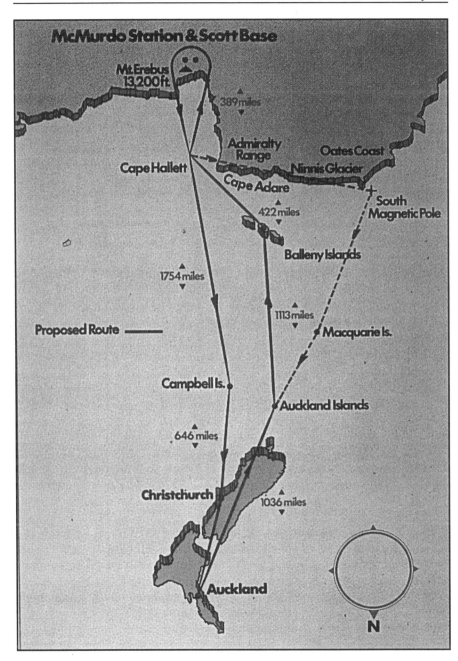

Figure 2.1.   Route of Antarctic flights
(Source: *The Erebus Papers*, Stuart Macfarlane, Avon Press, Auckland, New Zealand, 1991.)

> That mistake is directly attributable, not so much to the persons who made it, but to the incompetent administrative airline procedures which made the mistake possible.

Thus the official investigation, conducted fully in accordance with the provisions of ICAO Annex 13 (Accident Investigation), put the responsibility for the accident squarely upon the shoulders of the captain and crew, absolving the airline and the civil aviation administration from any responsibility – while the Commission of Inquiry completely absolved captain and crew, placing responsibility for the accident entirely on the airline and the civil aviation administration.

Failed defences, unsafe acts and error producing conditions

Introduction

The following review of the Erebus accident illustrates how accidents in modern technological systems result from the pernicious accumulation of delayed-action failures; from collective rather than individual errors. The discussion reviews some of the evidence, concentrating on the systemic failures and organizational deficiencies existing at the time of the accident. It is suggested that these bred latent failures which eventually combined with active failures – and a hostile operating environment – to breach system defences and cause the accident.

From a systemic risk management perspective, allocating blame for an accident has little merit. Given the fact that human error is a normal characteristic of human behaviour (Reason, 1990; Senders and Moray, 1991) the objective must be to identify those systemic failures which either foster and enable human error, or which fail to contain and negate its consequences. In this light, the findings of both the official Accident Investigation and the Commission of Inquiry could be considered to be correct, given that the flight crew failed in their role as the last line of defence. However, their performance equally reflected the behaviour which any flight crew could have been expected to display in similar circumstances, given their existing knowledge and their route briefing. This does not negate the fact that through their actions or inactions they triggered the accident. However, the surrounding aviation system itself manifested deficiencies and failures which enabled the flawed flight-crew performance.

Erebus – the flight

Ross Island, the general site for sight-seeing, was reportedly overcast with a cloud base of 2000 ft, with some light snow and a visibility of in excess of 40 miles. Clear areas some 75–100 miles northwest of McMurdo were also reported. Approximately 40 miles north of McMurdo the flight encountered sufficient breaks in the overcast to permit a visual descent to penetrate below the cloud base. The flight requested and obtained from the United States Navy Air Traffic Control Centre (Mac Centre) clearance to descend and proceed visually to McMurdo, with a request to keep Mac Centre advised of the aircraft's altitude. The automatic tracking feature (NAV track) of the Area Inertial Navigation System (AINS) was disconnected and a 'manual', orbiting descent in visual conditions to an altitude of 2000 ft was accomplished. The aircraft thereafter continued to 1500 ft to enhance passenger viewing of the landscape. When the aircraft levelled off, the AINS NAV track was reselected. Approximately three minutes later, the Ground Proximity Warning System (GPWS) gave a 500 ft warning. Fifteen seconds later, as the crew initiated a pull-up, the aircraft collided with the ice slope at the base of Mount Erebus (Figure 2.2).

Analysis

Only a limited number of failures from each level have been selected for review. The analysis is worked backwards from the accident to the latent organizational failures – in effect asking the question 'what cumulative acts of omission or commission enabled or facilitated the events which led to this accident?'. The reader may find it helpful to refer to Figures 2.2 and 2.3 from time to time.

Failed defences

The air transportation system has defences at many levels. On this occasion there were failures at all levels of defence:
- Neither the regulatory authority nor the operator ensured that the flight crew were provided with sufficient information regarding the peculiar Antarctic phenomenon of sector whiteout. Whiteout – which is *not* the reduction in visibility familiar to skiers – occurs under specific cloud and light conditions, *and leads to altered visual perception*. Whiteout undermines the protection provided by Visual Flight Rules (VFR) and renders white obstacles, even a few feet ahead of an

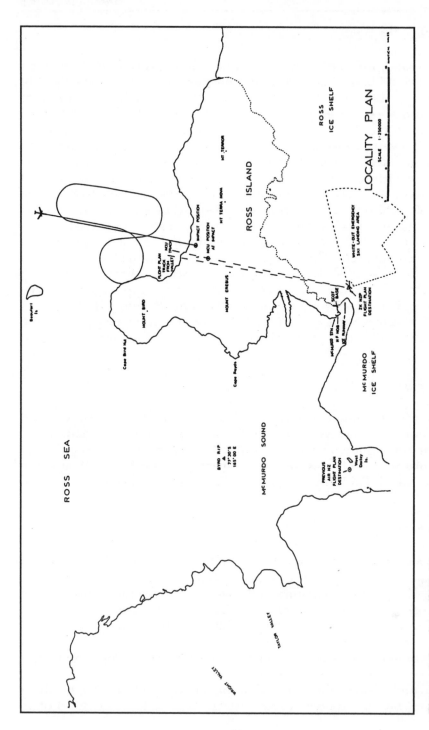

Figure 2.2. Locality plan
(Source: *Aircraft Accident Report No. 79–139.*)

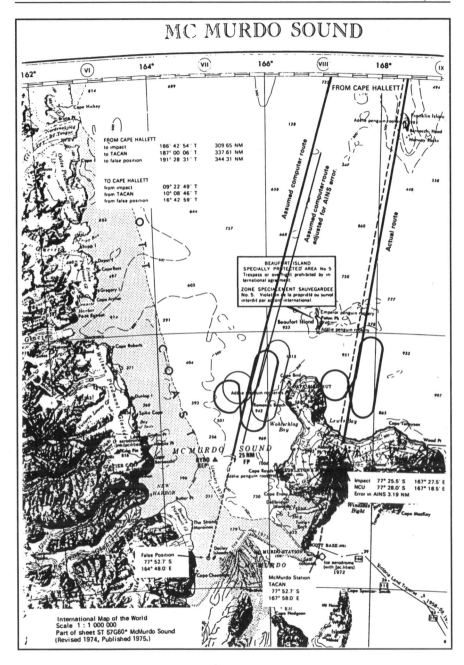

Figure 2.3. Actual and believed flight paths
(Source: Vette and Macdonald, 1983.)

41

aircraft, invisible – even if flying in nominal visibilities greater than 40 miles. Prior to undertaking Antarctic flights, pilots were admonished during route briefing to avoid snow showers and blowing snow. However no information was provided regarding the pitfalls of attempting visual separation from textureless white terrain while flying in overcast polar conditions. No alert, warning – or prohibition – was included in the route briefing materials regarding visual flight in overcast conditions. The crucial significance of this is immediately apparent when one realizes that it is impossible for an observer to determine that they are experiencing whiteout – the only fail-safe protection being to avoid the conditions in which it occurs.

- The computer flight plan presented to pilots during the Antarctic route briefing ran from Cape Hallet through the Ross Sea and the centre of McMurdo Sound, a flat expanse of ice 40 miles wide, to the Byrd reporting point, some 25 miles west of McMurdo Station, and thence to McMurdo. This closely followed the route used by routine military traffic, and had been followed by seven sight-seeing flights during the previous twelve months. This route placed Mt. Erebus 25 miles to the left of the flight path. However, on the night before the accident (about six hours before scheduled take-off time), the computer flight plan was modified by the operator's navigation section. The final way-point was now Williams Field, rather than Byrd reporting point. The new track, now displaced to the left, *lay directly across Mt. Erebus.* Neither the captain nor any flight crew member was notified of the change in the final destination way-point. When, after an orbiting visual descent, the crew reselected NAV track, the aircraft path was then connected to the (newly altered) computer flight plan, the crew undoubtedly believed they were heading over a flat expanse of snow towards McMurdo Sound. Instead, they were heading directly towards Mt. Erebus, hidden under the overcast by the insidious whiteout (Figures 2.2 and 2.3).

- The briefing provided to Antarctic crews included reference to two altitudes for the sight-seeing area – 16,000 ft to clear Mt. Erebus and 6000 ft, when within a defined sector around McMurdo. These safety altitudes provided protection from the high ground in each sector. After the accident, the alleged role and significance of these altitudes became the subject of heated controversy. Some flight crews from previous flights viewed them as minimum safe altitudes, independent of weather conditions. In that case the accident flight crew would have breached explicit restrictions by descending to 1500 ft. However, most other flight crews viewed them as cloud break altitudes and believed that visual descent to lower altitudes was permissible. The fact that pilots from previous Antarctic trips held conflicting views as to the role of these altitudes clearly illustrates that route briefing guidance was at least equivocal. Given some of the assertions after the accident, the absence of a clear prohibition on descent below a specified altitude, were that indeed intended, must be viewed as a failed defence (see Macfarlane, 1991, pp 325–6, for details). It is manifestly the case that unclear lines of communication help foster flawed decision-making by operational personnel.

- Planning for Antarctic flights stipulated that before being assigned as pilot-in-command, captains would undertake one familiarization flight under the supervision of another captain who had already visited Antarctica. Such sensible precautions are normal in airline operations, and it was here intended to ensure that at least one flight deck crew member was familiar with the alien Antarctic environment. While flight crew rostering of earlier flights observed this requirement, it was eventually discontinued. The justification for this decision was that the route briefing fulfilled all familiarization requirements. In consequence, neither the captain of the accident flight, nor either of the first officers, had been to the Antarctic before.

- The flight crew had been informed that the McMurdo NDB (a navigational beacon) was unavailable. This left the distance measuring portion of a military navigation aid as the only ground-based navigational guidance available. During route briefing of the accident crew, the briefing officer experienced considerable difficulty in ascertaining the real status of the NDB. In fact, the NDB was transmitting, but it had been formally withdrawn from use, and the U.S. Navy had decided that it should continue transmitting until it failed. Thus, for all practical purposes, the NDB was available as a further supporting aid. Deficient communications prevented the flight crew from being aware of this fact.

- In accordance with their regulations, the U.S. Navy had informed the State's civil aviation authority that Air Traffic Control (ATC) support would only be available on a limited basis, and restricted to providing advisory information. Whether this state of affairs was fully appreciated by flight crews is open to debate, as Antarctic crews were briefed to deal with McMurdo ATC as they would do with any other ATC unit, despite the

43

prevailing restrictions. ATC is one of the aviation system's primary defences. Any limitations to ATC services debilitate system defences, and even more so if such limitations are unknown to its users.

Unsafe acts

During the final stages of the flight, the crew arguably committed several unsafe acts, leading directly to the accident:

- They descended below two safety altitudes without definitively ascertaining their position.
- They did not use the available facilities at Williams Field to help accomplish their descent.
- They attempted visual separation from terrain in whiteout conditions.
- They continued to attempt visual separation from terrain while flying towards an area which, according to the CVR transcript, may have occasioned doubts as to the possibility of maintaining visual conditions.
- They did not react to the fact that, in spite of the proximity to McMurdo, they were unable to establish VHF radio contact. This might have suggested some obstacle (such as high ground) between their position and McMurdo.
- They did not use the AINS information to cross-check the track expected to take them to the centre of McMurdo Sound and to the Byrd reporting point.

Aside from these issues, there is one unsafe act in particular which would not be explicable unless considered within proper context: how could an experienced captain and his equally experienced crew misperceive the dangers and pitfalls associated with low-level polar flying in the conditions existing on the day of the accident? The captain was held in high esteem by managers and peers alike, who regarded him as a conscientious and conservative professional. His pre-flight preparation testifies to this fact. His team mates were, by all accounts, equally experienced, conscientious and knowledgeable. The available evidence suggests that the crew performed as an intact team, working in effective co-ordination until the last second.

Several factors may have prompted the crew to descend to a low altitude and continue flying under the prevailing conditions:

- The flight crew were completely unaware of the dangers of sector whiteout. They clearly assumed that, since they had excellent lateral visibility (in excess of 40 miles), the apparent vastness of white ahead of them represented an unobstructed flight path suitable for visual flight. They may have assumed that, in line with the route briefing they had received, so long as they stayed away from snow showers or blowing snow, they would be safe from visibility impairing conditions.

- The flight crew knew that previous Antarctic flights had descended to low altitudes in visual conditions and had conducted their sight-seeing safely, maintaining visual separation from detectable obstacles.
- The flight crew believed that their track ran down the centre of McMurdo Sound, with Mt. Erebus well to the left; that this was so is endorsed by the fact that, after level off at 1500ft, they re-engaged the NAV track feature on the AINS. Their belief may have been further reinforced by the (erroneous) identification of landmarks – which unfortunately were themselves fully coincident with the believed track (these are outlined in Vette and Macdonald, 1983).

Error-producing conditions

The error-producing conditions which led the flight crew to make flawed decisions and to commit unsafe acts were a by-product of failures in the aviation system's defences. These include critical gaps in the regulatory and operational safety net, ambiguity in the Antarctic operating procedures, inadequate information about the nature of sector whiteout and Antarctic inexperience – along with associated route briefing and training deficiencies.

The operator's route briefing was approved by the regulatory authority. It included most of the information required to familiarize flight crews with Antarctic flights, and included a flight simulator detail. The briefing represented a broad framework which should have ensured the safety of these flights. There were, however, omissions and error-producing ambiguities in both the route and simulator briefings. These remained undetected until after the accident.

The Antarctic flights were also predicated on a set of requirements, some of which were difficult to reconcile with the sight-seeing intentions of the operator. In practice the mismatch between these requirements and situational contingencies induced deviations – that themselves were aimed at achieving the sight-seeing objective. The two altitudes quoted in the briefing constitute one example. They were not the best altitudes for sight-seeing, and previous Antarctic flights had descended below both altitudes. If they were truly intended as *minimum altitudes*, such descents constituted a deviation from intended procedures. However, low-level sight-seeing was a matter of public knowledge, and had not been challenged by either the operator or the authority. Even if the altitudes had been intended as a minimum, this acquiescence became an error-inducing condition; it would perforce have influenced the accident crew's belief that they were acting in accordance with acceptable precedent.

The route briefing included information regarding an instrument let-down over the McMurdo NDB, using a U.S. Navy-published instrument procedure. This was available to help acquire visual conditions for the sight-seeing. After the NDB was formally decommissioned the only way these flights could descend to sight-seeing altitudes was using Visual Flight Rules, and the route briefing was amended accordingly. Although seemingly simple and straightforward, this amendment contained error-inducing potential. It must be borne in mind that this was a commercial flight, dispatched under the strict provisions of Instrument Flight Rules (IFR) to operate under augmented Visual Flight Rules (VFR) in the vicinity of a nominal destination point. However, it is important to understand that VFR confer considerable autonomy and flexibility on the pilot, provided that the Visual Flight Rules themselves are not infringed.

Therefore, when approaching the McMurdo overcast, with no navigation aids to execute the let-down, and upon finding a large break in the clouds which allowed for visual descent, the flight crew resorted to the autonomy provided by VFR to accomplish their intended objective. This decision was correct within the framework provided by the briefing, the knowledge existing at the time of the accident, the experience obtained from previous Antarctic flights and the operational behaviours all these circumstances generated. A further reason why visual descents were perceived to be a 'normal constituent' of Antarctic flights was the fact that there was an alternative (bad weather) flight routing for which only a visual descent was possible.

Air Traffic Control remained the last line of defence. However, the decision to descend visually was probably reinforced by the 'clearance' received from McMurdo, which in no way challenged the descent or questioned its appropriateness. With the benefit of hindsight, the decision to descend turned out to be an error. The explanation for the error must be sought, however, far and away from McMurdo Sound and Mt. Erebus.

Latent organizational failures

Chapter 1 introduced the idea of organizational pathogens, to be found in all high-technology systems. The thesis advanced here is that such pathogens are invariably the true root causes of accidents within high-technology industries, and that the generating conditions can, for example, include:

- A lack of top-level management safety commitment or focus.
- The creation of conflicts between production and safety goals.
- Poor planning, communications, monitoring, control or supervision.
- Organizational deficiencies leading to blurred safety and administrative responsibilities.
- Reliance on oral communication of critical information in the absence of written records.
- Deficiencies in training.
- Poor maintenance management or control.
- Monitoring failures by regulatory or safety agencies.

A number of these played a part in the chain of events which led to the Erebus accident:

- There is no evidence of a lack of management commitment to safety on the operator's part; indeed, the operator's commitment towards safety was obvious. There is, however, evidence which suggests that this commitment was not – within the context of this accident – successfully carried through into practice. For example, although it was widely known that low altitude flying of previous flights had violated what were purported to be minimum safe operating altitudes, there was no reaction by the operator to reports of these 'violations'. The assessment of Justice Hidden (1989), when reviewing management commitment to safety in the Clapham Junction railway accident, suggests a possible parallel:

  > The evidence therefore showed the sincerity of the concern for safety. Sadly, however, it also showed the reality of the failure to carry that concern through into action. It has been said that a concern for safety which is sincerely held and expressly repeated but, nevertheless, is not carried through into action, is as much protection from danger as no concern at all.

47

- There were indications of blurred organizational safety responsibilities. Various operational managers were aware of descents to low altitudes. However, each of them perceived that taking action – in response to possible violations of operational restrictions – was not their individual responsibility. Another indication of blurred responsibilities was evidence provided by the company safety officer; he asserted that, until the accident, he ignored the fact that the track of the flights at the original planning stage ran directly over the top of an active volcano.

- Crucially, it would appear in retrospect that the exact physical and perceptual nature of sector whiteout was not fully understood by those planning and approving the series of Antarctic flights, at least until well after the accident.

- At the time of the accident, the production goals of the Antarctic flights had been fully accomplished. The flights were an unqualified success. The conflict between these production goals and the associated safety goals gives pause for thought, in particular when considering the 'tools' available to the crews to discharge their duties. This became critical after decommissioning of the McMurdo NDB, when a visual descent became the only means to achieve the flight's production goals. Failing to resolve the conflict of scheduling commercial flights to conduct sight-seeing within a precise sector defined by two navigation aids (NDB/TACAN) became significant when no changes were made to operational procedures in response to the new state of affairs.

- Dropping the practice of rostering two captains and one first officer in favour of one captain and two first officers reduced the safety margin associated with the redundancy this provided. This was further reduced when the requirement for familiarization flights for captains was discontinued.

- The operator had received a communication, directed from the U.S. Navy Support Force in Antarctica to the civil aviation authority, indicating that limited Search And Rescue (SAR) capability existed over land, and very little over water. Although SAR issues were not relevant to the accident, the question of the conflict between production and safety goals again arises here. It is a matter of conjecture what the consequences could have been, in terms of loss of life due to exposure, had the accident been survivable.

- The coordinates of the final way-point were changed by the operator's navigation section within a few hours of departure. That neither Dispatch, nor the flight crew, were notified suggests failures of procedure and communication. The doubts generated by the lack of clarity as to minimum altitudes during route briefing stand as a further example. Doubts as to the actual track between Cape Hallet and McMurdo, which were made evident during the inquiry, are yet another example. These doubts represent a recurring latent failure: the absence of clearly stipulated management guidance to the relevant operational personnel.

- The operator had conducted several rounds of consultations with organizations operating in the Antarctic prior to determining the feasibility of the Antarctic flights. Once a decision to proceed with the flights was made, and the operation approved by the civil aviation authority, no further contacts were made with these organizations to obtain feedback – based on their experience in Antarctic operations – about the feasibility and soundness of the procedures selected. For example, no contact was made with the U.S. Navy to discuss with McMurdo ATC how the flights would operate, their intended route(s), the descent sector selected, and so forth. As a consequence, it was determined after the accident that the sector chosen for instrument let-down was unsuitable, given the McMurdo ATC facilities. The available evidence thus supports the notion that there was room for improvement in the planning of the Antarctic flights.

- Inadequate control and supervision are best illustrated by the controversy surrounding descents to low altitudes. It is similarly evidenced by the fact that, over a twelve-month period, seven flights operated into Antarctica with an error in the destination way-point computer flight plan coordinates before that error was detected. Furthermore, at the time of the accident, the route briefings were based upon a computer flight plan containing these erroneous coordinates, and with such coordinates underlined or highlighted.

- The route and simulator briefings were the appropriate means by which to convey management's intentions and goals to pilots. These would include both production *and* safety goals. Briefings, if they are to accomplish their objectives, must provide precise information and guidance, rather than oblique references. All the relevant operational and safety information should have been part of the Antarctic route briefing materials. It was not. Some of the major omissions and mistakes have already been discussed: incomplete information in regard to whiteout; ambiguous information about what were effectively the

49

minimum safe altitudes; misleading information about the intended route of flight; and ambiguous information about the nature and extent of the air traffic services provided. That the briefing failed to produce the desired behaviours attests to its failure, and suggests the consequent existence of allied training deficiencies.

- Finally, regulatory failures were evident in this accident, since most of the latent organizational failures could have been avoided by diligently applying existing regulations. The Antarctic operation had been scrutinized and approved by the relevant authorities at the very outset, when there was a failure to notice the incomplete information provided to flight crews. Approving a route which, at the planning stage, passed directly over an active volcano can be regarded as another regulatory failure. As the operations evolved and departed from those originally approved (dropping the requirement for familiarization flights; decommissioning of the NDB, etc.), the regulatory authority required no changes. A rather passive attitude was also the answer to the much-publicized low-altitude sight-seeing flights. If they indeed represented violations of intended restrictions, it was the regulatory authority's role to adopt an active stance on the matter.

Conclusion

The wisdom of hindsight: lessons for foresight

Despite the many caveats expressed above, some readers may regard the analysis as an indictment of the airline involved and, perhaps, also of the State's operating practices. These readers might reassure themselves that their organization is immune from the practices and failures discussed here – a syndrome typified by statements such as 'that kind of thing certainly would not happen here, we've got "X" or "Y" protections in place'. In so doing, not only would they be wrong, but they would also be missing the entire point of this analysis. Just as error is a normal component of human behaviour so, equally, every organization harbours latent failures. Just as human error is inevitable, so too are fallible management decisions. Just as active failures – occasioned by human error or equipment failure – are inevitable, so too are mistakes bred by latent failures.

It is easy to be clever in retrospect. It is also relatively easy to rework the findings of official investigations from different methodological standpoints and present the relevant events in a new light. The real challenge is to provide a coherent and productive analysis which is also true to the events in question. The preceding review of the Erebus accident is intended to demonstrate the analytic value of the model outlined in Chapter 1. Hopefully it will also help the reader draw key preventative lessons and assess how accident investigation might be further improved. The foregoing analysis is therefore intended to demonstrate the applicability of a new way of looking at accidents and human error. The overall objective is to promote the case of accident prevention, and not to further dissect the accident at Mount Erebus. The intentions are forward-looking – for reasons best summed up in Santayana's comment that 'Those who cannot remember the past are condemned to repeat it'.

Proactive and reactive organizations

The need for organizational methods to help identify and contain the consequences of fallible management decisions will be obvious from the foregoing analysis. In the 'system safety' scheme of things, immediate failures on the part of individuals can be deemed irrelevant for all practical purposes, save the identification of essential changes to organizational systems. The key to proactive safety management lies in identifying latent failures and remedying them before their consequences are visited upon the organization.

There are two basic requirements for effective risk management programmes. One is suitable organizational structures and processes. The other is to ensure adequate feedback regarding the quality of those organizational processes. Effective risk management systems must necessarily have many layers and employ multiple techniques. In an adequately specified and structured system, the priority will be to ensure that feedback is relevant, valid, timely and accurate, so that prompt and enduring interventions may be initiated. The fundamental requirements are therefore structural and cultural, given that effective risk management structures can only function effectively in the presence of an organizational sub-culture which endorses and promotes feedback and remediation.

Organizations which are able and willing to respond promptly to feedback and modify their relationship to both the internal and external operations environments have been described as 'generative' (Westrum, 1992); such organizations can be contrasted with closed or 'pathological' organizations (Westrum, *op cit.*), complete with their blame-culture, the absence of feedback and employees who are adept at hiding errors and playing the politics of denial. In other words, 'pathological' organizations are prone to elevated risk, through the cultivation of latent errors.

Management are responsible for establishing the necessary organizational structures and ensuring that appropriate operational policies are in place (Bruggink, 1985). Management are also responsible for ensuring that those policies are effectively translated into action (Wiener, 1993; Degani and Wiener, 1994).

Individuals and organizations

The author of a dissenting statement to a recent NTSB report (NTSB, 1991, p.44) asserted that *people* have accidents, not *organizations* (or 'agencies'):

> I... disagree with the notion that agencies cause accidents. Failures of people and failures of equipment cause accidents. Shifting the cause from people to agencies blurs and diffuses the individual accountability that I believe is critically important in the operation and maintenance of the transportation system.

The author is clearly concerned at the consequences of diluting the role played by the concept of individual responsibility. There is a dilemma here which needs to be briefly considered. On the one hand, we must recognize the importance of individual accountability, while on the other we must recognize that front-line personnel do not act in a manner which is independent of company working custom and practice – or 'organizational sub-culture'. In respect of the role played by organizational working 'realities', it may be appropriate to employ a *substitution test*. This merely involves mentally substituting another actor from the same operational background into the circumstances of an accident or incident, and asking the question, 'In the light of existing knowledge and how events unfolded sequentially, is it probable that this new individual would have behaved any differently?' (Johnston, 1991).

If the answer is no, it is suggested that apportioning blame will have no material role to play, other than to hide systemic deficiencies and to blame one of the victims. By way of example, in the case of the Zeebrugge ferry disaster it is clear that the behaviour of the captain and crew on the occasion of the accident differed in no material way from that which had occurred on numerous previous occasions (Sheen, 1987). This accident was precipitated by an individual who simply fell asleep – the single exceptional event. As the subsequent investigation was to conclude, this was a disaster waiting to happen, given that the organization was infected from top to bottom by the disease of sloppiness. All the central actors turned out to be victims of the circumstances in which they found themselves, given the lack of top management support and the fact that each layer of safety protection had previously been subverted or rendered ineffective.

On the other hand, this approach does not automatically mean we must avoid apportioning *responsibility* for acts or omissions. Indeed, it probably will be the case that the most effective preventive action will equally centre on ensuring that managers and front-line workers fully understand – and are willing, and able, to discharge – those very responsibilities. However, the reasons why management and workers fail to act in accordance with their formal responsibilities is itself a fundamental issue which organizations must successfully confront if latent organizational pathogens are to be eradicated.

Coda

The investigations and associated literature relating to the Erebus accident were mainly produced ten years before the Dryden accident (discussed in the following chapter). The Erebus findings generated violent local controversy, but remained inconspicuously shelved until recently. However, in the context of the approach taken in this book, the *Report of the Commission of Inquiry* was ten years ahead of its time. After all, Chernobyl, Bhopal, Piper Alpha, Clapham Junction, King's Cross and other major high-technology system catastrophes had yet to happen. They need not have happened. In retrospect, if the aviation community – and the safety community at large – had grasped the central message from Antarctica and applied its preventative lessons, Chernobyl, Bhopal, Piper Alpha, Clapham Junction, King's Cross and, certainly, the Dryden accidents would have been avoided.

It is also worthy of note that the key ground-breaking accident reports of recent times benefited directly from the involvement of Judicial Officers, charged with wide-ranging investigatory powers. Each of these Judicial Officers was prepared to consider the relevant events within their surrounding context, and thus to escape the narrow perspective deriving from the formal accountability of front-line workers. It is an interesting paradox that these important initiatives were taken by Judicial Officers, and that few professional accident investigators have, for various reasons, heretofore felt free to show similar initiative.

This chapter aimed to demonstrate the merits of the model outlined in the opening chapter. These merits include:

(1) An analytical model which captures the salient features of human and organizational error.
(2) A model which guides both reactive and proactive approaches to accident prevention.
(3) A suitable framework to facilitate the further improvement of Human Factors accident and incident investigation.
(4) Guidance for management as to the organizational characteristics of effective safety practice.

Acknowledgements

We acknowledge the assistance of the Gordon A. Vette, Flight Safety Foundation, Auckland, New Zealand, with Figures 2.1 and 2.3. Figure 2.2 is from *Aircraft Accident Report No. 79-139*, Office of Air Accidents Investigation, Wellington, New Zealand.

References

Alicke, M. D. (1992) Culpable causation. *Journal of Personality and Social Psychology*, **63**: 368–378.
Aircraft Accident Report No. 79-139; Air New Zealand McDonnell-Douglas DC10-30 ZK-NZP Ross Island, Antarctica, 28 November 1979 (1980) Office of Air Accidents Investigation, Ministry of Transport, Wellington.
Andersen, N. S. and Wreathall J. (1990) A structure of influences on management and organisational factors on unsafe acts at the job performer level. *Proceedings of the Human Factors Society, 34th Annual Meeting*, **34**, 881–884.
Barnaby, K. C. (1968) *Some Ship Disasters and their Causes*. London: Hutchinson.
Bruggink, G. M. (1985) Uncovering the policy factor in accidents. *Air Line Pilot*, May 1985.
Degani, A. and Wiener, E. L. (1994) *Philosophy, Policies, Procedures, and Practices: The Four "P"s of Flight-Deck Operations*. In: Johnston, A. N., McDonald, N. J. and Fuller, R. G. (Eds) *Aviation Psychology in Practice*. London: Ashgate.
Denning, M. R. (1978) Alidair v Taylor. London: *Industrial Court Reports* 445. OUP.
Einhorn, H. J. and Hogarth, R. M. (1986), Judging Probable Cause. *Psychological Bulletin*, **99**(1), 3–19.
Fischoff, B. (1994) Acceptable risk: a conceptual approach. *Risk: Health, Safety & Environment*, **5**(1), 1–28.
Grose, V. L. (1987) *Managing Risk: Systematic Loss Prevention for Executives*. Englewood Cliffs; NJ: Prentice Hall.
Hidden, A. (1989) *Investigation into the Clapham Junction Railway Accident*. Department of Transport, London: HMSO.
Hood, C., Jones, D. K., Pidgeon, N. F., Turner, B. A. and Gibson, R. (1992) Risk management. In: *Risk: Analysis, perception and management*. London: The Royal Society.
International Civil Aviation Organization (1994) Human Factors, Management and

Organisation. ICAO *Human Factors Digest Number 10.* Montreal.

Johnston, A. N. (1985, December) Erebus. *The Log,* 16–18.

Johnston, A. N. (1989) A human performance re-interpretation of factors contributing to an aviation accident. *Proceedings of the 5th International Symposium on Aviation Psychology,* Columbus, OH 848–853.

Johnston, A. N. (1991) Organisational factors in human factors accident investigation. *Proceedings of the 6th International Symposium on Aviation Psychology.* Columbus, OH 668–673.

Johnston, A. N. (in press) Blame, punishment and risk management. In: Hood, C., Jones, D., Pidgeon, N. and Turner, B. (Eds) *Accident and Design.* London: University College Press.

Macfarlane, S. (1991) *The Erebus Papers.* Auckland, New Zealand: Avon Press.

Mackie, J. (1974) *The Cement of the Universe: A Study of Causation.* Oxford: OUP.

Maurino, D. (1992) Corporate culture imposes significant influence on safety. *International Civil Aviation Organization Journal,* April 1992.

Moshansky, Mr. Justice (1992) *Commission of Inquiry into the Air Ontario Crash at Dryden, Ontario.* Final Report, Ottawa, Canada.

National Transportation Safety Board (1991) *Aircraft Accident Report: Runway collision of Eastern Airlines Boeing 727, Flight 111 and Epps Air Service Beechcraft King Air A100 Atlanta Hartsfield International Airport, Atlanta Georgia, January 19, 1990.* Washington, DC.

National Transportation Safety Board (1992) *Aircraft Accident Report: Britt Airways, Inc., d/b/a Continental Express Flight 2574 In-flight structural breakup EMB-120RT, N33701 Eagle Lake, Texas, September 11, 1991.* Washington, DC.

Reason, J. (1990) *Human Error.* Cambridge: Cambridge University Press.

Senders, J. W. and Moray, N. P. (1991) *Human Error.* Hillsdale, NJ: Laurence Erlbaum.

Sheen, Mr Justice (1987) *MV Herald of Free Enterprise. Report of Court No. 8074 Formal Investigation.* London: Department of Transport.

Trigg, R. (1985). *Understanding Social Science.* Oxford: Blackwell.

Trusted, J. (1981) *Theories and Facts.* Open University Course Units, U202. Milton Keynes: Open University Press.

Westrum, R. (1992) Cultures with requisite imagination. *NATO Verification and Validation Conference,* Vimeiro, Portugal.

Wiener, E. L. (1993). *Intervention strategies for the management of human error.* NASA CR-4547. Moffett Field, CA: NASA-Ames Research Center.

Erebus bibliography

Reports

Report of the Royal Commission to Inquire into the Crash on Mount Erebus, Antarctica, of a DC10 Aircraft Operated by Air New Zealand Limited (1981) Wellington.

Books

Guy, M. (1980) *Whiteout.* Alister Taylor.

Hickson, K. (1980) *Flight 901 to Erebus.* Whitcoulls.

Mahon, P. (1984) *Verdict on Erebus.* Collins.

Vette, G. with Macdonald, J. (1983) *Impact Erebus*. Auckland, New Zealand: Hodder & Stoughton.

Court Cases

High Court
Unreported, Mr. Justice Speight: Auckland, A 482/81, 15 July 1981.

Court of Appeal
Re Erebus Royal Commission; Air New Zealand v Mahon [1981], Volume 1, New Zealand Law Reports, p. 614.
Re Erebus Royal Commission; Air New Zealand v Mahon (No 2) [1981], Volume 1, New Zealand Law Reports, p. 618.
Re Royal Commission on Thomas Case [1982], Volume 1, New Zealand Law Reports, p. 252.

Judicial Committee of the Privy Council (House of Lords)
Privy Council Appeal Number 12. (1983).
Re Erebus Royal Commission; Air New Zealand v Mahon [1983] New Zealand Law Reports, p. 662.

3 Pathogens in the snow: the crash of Flight 1363

There are ill discoverers who think there is no land when they can see nothing but sea.

Samuel Johnson

Introduction

At 12:09 pm Central Standard Time (CST) on Friday, March 10, 1989, Capt. George C. Morwood, age 52, a 24,000-hour flight time veteran, including 673 hours' experience in jets, advanced the throttles of C-FONF, Air Ontario Flight 1363, a Fokker F28 1000, initiating the take-off roll from runway 29 at the small provincial airport of Dryden, Ontario, Canada. His actions were scrutinized by First Officer Keith Mills, age 35 and a veteran in his own right, with more than 10,000 hours' experience including 3500 hours in jets. It is possible that the start of the take-off roll was accompanied with a sigh of relief by the conscientious, 'fly-by-the-book' captain. So far, March 10 had been a frustrating day in which weather, heavier than forecasted passenger bookings and the less-than-optimum operational status of C-FONF had combined to produce numerous delays. This bothered Capt. Morwood, a professional known by his commitment to on-time performance and concern for his passengers. But the end of the day – and the frustrations – was around the corner, a mere 45 minutes away, in Winnipeg, Manitoba.

From the vantage point of a neutral observer, there were obvious

57

reasons which should have induced Capt. Morwood and First Officer Mills to seriously reconsider the decision to take off. Equally obvious should have been the need to ponder the adoption of some precautions during the well-rehearsed, routine actions of pre-departure checks and procedures. Both flight attendants and many passengers, including two active airline captains deadheading back to Winnipeg, had noted with concern an accumulation of snow over the wings during taxi out. Such concerns were, however, never transmitted to the flight crew. In any event, beyond all possible speculation, the fact remains that in Capt. Morwood's mind, under the existing circumstances, based on his knowledge and understanding of the situation and his perception of the surrounding events, he was making the right decision by spooling up the two Rolls-Royce Speys to take off thrust. With the benefit of hindsight, he was not.

Less than one kilometre from the end of runway 29, Flight 1363 became a mound of smouldering metal and the death trap for 21 passengers and three crew members, including Capt. Morwood, First Officer Mills and Flight Attendant Katherine Say. A lethal combination of wet snow and ice had accumulated over the wings of C-FONF and inhibited their lifting capabilities, leading to an unrecoverable stall after lift off. The obvious question, which eluded immediate answer, was *why*, especially in view of the flight crew's experience in cold weather operations. Why did such experienced crew, exposed throughout their professional careers a thousand-and-one times to scenarios similar to that of Dryden, ignore all the tell-tale indicators presented to them on the day of the accident? Why did two normal, healthy, competent and properly certificated individuals allow their well-equipped, state-of-the-art machine to find its way to its own – and their – destruction? Why do humans make such obvious and damaging errors? The answers to these questions were not to be found 'kicking tin' at Dryden, Ontario, but far and away, in corporate suites, training departments and at the desks of regulators.

What the crew of Flight 1363 could have never possibly known was that their ill-fated decision to take off from Dryden without de-icing the wings of C-FONF on March 10, 1989, would be the origin of one of the most extensive and all-encompassing investigations in the history of aviation, similar in scope but widely exceeding the resources invested by The Honourable Mr. Justice Mahon during his inquiry into the Erebus crash, discussed in Chapter 2. It was clear *what* had happened: the aircraft had crashed because its wings were covered with ice and snow. It would seem equally clear why it had happened – at least, from the perspective afforded by conventional knowledge: the pilots had made a wrong decision. If it was snowing and icing conditions prevailed, and the pilots elected to take off without de-icing the wings, then the accident *must* have been caused by pilot error.

The evidence was so overwhelmingly obvious that the investigation might have been closed within a few weeks of the crash, had conventional knowledge been applied. However, this was not to be the case. It was evident to many within the Canadian aviation community that there was more to the story behind the crash of Flight 1363.

On March 29, 1989, a Commission of Inquiry was formed and The Honourable Mr. Justice Virgil P. Moshansky was appointed Commissioner. Without hesitation, Mr. Justice Moshansky discarded what was as obvious as inconsequential – the notion of pilot error – and over a period of 20 months literally tore apart the Canadian aviation system, diligently advised by a multi-disciplinary team of investigators, safety officials, Human Factors practitioners and researchers, pilots, engineers and regulators, trying to find the answers behind the flight crew's decision to take off.

Mr. Justice Moshansky interpreted his mandate in the broadest sense:

> The mandate of this Commission was to investigate a specific air crash and to make recommendations in the interests of aviation safety. In carrying out this mandate, it was necessary to conduct a critical analysis of the aircraft crew, of Air Ontario, of Transport Canada and of the environment in which these elements interacted...I have adopted a system-analysis approach, with emphasis on an examination of human performance. (Moshansky, 1992, pp xxv)

After a total of 168 hearing days, 166 witnesses, 168 volumes of transcript totalling some 34,000 pages and contents of documentary exhibits totalling more than 177,000 pages, the four-volume, 1700-plus page *Final Report of the Commission of Inquiry into the Air Ontario Crash at Dryden, Ontario* was made public, in late March 1992. The Dryden Report, as it has become widely known, is an accomplished exercise in accident investigation and prevention from the systemic as well as from the organizational perspective. The value of the Report lies beyond its *post mortem* lessons, that is, the accurate reconstruction of the circumstances surrounding the event and the proposal of remedial measures aimed at avoiding similar operational personnel errors and thereby accidents similar to Dryden in the future. What makes the Report a turning point in contemporary aviation safety are its *premortem* lessons. Those in charge of managing aviation operations – be they trainers, regulators or flight managers – can find in the Report both a guide to determine the state of health of their operations and a 'how to' improve 'sick' or weak areas.

The Report leaves no room about where to look to improve safety within the Canadian aviation system. Its lessons are unquestionably relevant to any aviation system of today:

> The public hearings... disclosed numerous safety-related deficiencies and failings within the carrier, Air Ontario, specifically; within the aviation industry generally; and in the regulatory domain of Transport Canada. These shortcomings, their causes, and their relationships to the accident at Dryden must be closely scrutinized... (pp xxi) The aircraft crew members must contend with the total operating environment of a given flight and any constraints placed upon them by their aircraft, their air carrier, the immediate operational infrastructure, and the regulator...Had the system operated effectively, each of the factors might have been identified and corrected before it took on significance. It will be shown that this accident was the result of a failure in the air transportation system. (Moshansky, 1992, pp 5–6)

The Commission did not produce any probable cause statement. It did not try to encapsulate – in the simplistic manner characteristic of many investigation agencies – the very complex processes involved in an aviation accident. Instead, the Commission produced 191 recommendations which, blended with the body of the Report, permit positive identification of the latent failures within the Canadian aviation system. Such failures harbour the potential to generate accidents like Dryden, and accidents of any kind; they are failures that must be addressed by positive action from high-level, strategic decision makers and managers. The Report also identifies the preconditions which may foster active failures of operational personnel in contexts similar to Dryden and elsewhere.

The investigation of the crash of Air Ontario 1363 represents one of the first large-scale applications of a systemic, organizational approach to the investigation of an aviation accident. The balance of this chapter presents an abridged description of the events and their subsequent analysis from the perspective provided by Chapter 1.

The events

Air Ontario Flight 1362 departed Winnipeg, Manitoba, at 7:49 am CST, 24 minutes behind its scheduled departure time, a delay attributed to Capt. Morwood's request to de-ice the aircraft. The day's flying schedule for the crew consisted of a Winnipeg to Thunder Bay round trip, with intermediate stops at Dryden (flights 1362/1363), followed

Figure 3.1. Route of Flight 1363
(Source: *Final Report of the Commission of Inquiry into the Air Ontario Crash at Dryden, Ontario.*)

by another Winnipeg to Thunder Bay round trip without the Dryden intermediate stop (flights 1364/1365). Weather was poor in the region, and forecast to deteriorate below landing limits at Thunder Bay and Winnipeg. This dictated the need for distant alternatives and an increase in fuel requirements. Thus, the alternative selected for the flight from Dryden to Winnipeg was Sault Ste Marie instead of Thunder Bay. On the other hand, heavy passenger loads from Thunder Bay to Winnipeg due to the start of the spring school break required a reduced fuel uplift at Thunder Bay and created the necessity of fuelling at Dryden (Figure 3.1).

When Capt. Morwood reviewed the operational status of C-FONF before departing Winnipeg, he verified that among other maintenance deferred defects, the Auxiliary Power Unit (APU) remained unserviceable: it had been malfunctioning for the previous five days. The operational implications of this defect were multiple. The engines had to be started from an external power unit, or one engine had to be kept running to 'cross-start' the other engine. If both engines were shut down at a station where no external power unit was available, C-FONF would be stranded until the APU was fixed[1] or an external power unit was made available. Although one recourse could have been to have another turboprop or turbojet provide compressed air to start the engines, in what is known as 'buddy start', Air Ontario did not have the equipment readily available to perform such a start in the F-28. There was no external power source at Dryden, and therefore an engine would have to be kept running. The manufacturer, Fokker, and Air Ontario strictly prohibited de-icing with either engine running (although many European carriers did de-ice with engines running). Furthermore, this situation imposed the need for a 'hot refuelling' (refuelling with one engine running).

After de-icing the aircraft to clean some frost deposited overnight on the wings, Flight 1362 departed Winnipeg on an uneventful flight to Dryden, where it arrived 13 minutes late. Capt. Morwood discussed the operation into Thunder Bay by telephone with Air Ontario Systems Operations Control (SOC) located in London, Ontario. The weather at Thunder Bay had dropped below landing limits. Based on an improvement predicted in the forecast, and since the ground hold at Dryden would have had to be performed with one engine running, it was agreed that Flight 1362 would depart Dryden in the hope the forecasted improvement would materialize. While en route, the weather improved and the aircraft eventually arrived in Thunder Bay some 20 minutes behind schedule.

Because the original flight release from Thunder Bay to Dryden prepared by Air Ontario SOC had not been updated, ten passengers were added to Flight 1363 after it had been fuelled. Now overweight for take-off, Capt. Morwood elected to off-load the ten passengers and their baggage. However, the Air Ontario SOC duty manager overrode the Captain's decision and elected to accomplish the weight reduction by off-loading fuel, a total of about 2823 pounds. The defuelling caused an additional 35-minute delay in the departure of Flight 1363 from Thunder Bay and increased the 'hot refuelling' time at Dryden. Many passengers who had connecting flights in Winnipeg voiced their concerns to the cabin crew, concerns which eventually made their way to Capt. Morwood and First Officer Mills. Evidence from eyewitnesses disclosed that these events changed the good-spirited mood which the

---

[1]Technically speaking, this is incorrect. The APU fire detection system was inoperative, but the APU itself would have started and delivered pneumatic and electrical power as required. What is important here is the perception of Morwood and Mills, since all evidence suggests that their belief was that the APU was inoperative.

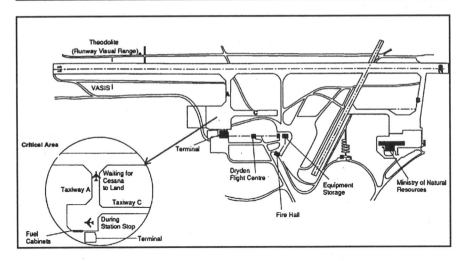

Figure 3.2. Layout of Dryden Municipal Airport
(Source: *Final Report of the Commission of Inquiry into the Air Ontario Crash at Dryden, Ontario.*)

flight crew had showed earlier during their stay in Thunder Bay.

At 11:55 am Eastern Standard Time (EST), with a full load of passengers, Flight 1363 departed Thunder Bay about one hour behind schedule. As the flight landed in Dryden at 11:39 am CST, still with the hour-delay, the first snow flakes started falling (Figure 3.2). While the aircraft taxied towards the terminal '...the snow was light and the weather gloomy and overcast' (Moshansky, 1992, p 49).

Eight passengers deplaned at Dryden and seven passengers, two of them infants, boarded the aircraft. In the meantime, the 'hot refuelling' process started, with the passengers remaining on board. Hot refuelling with passengers on board is considered to be an unsafe practice, quite difficult to reconcile with Capt. Morwood's style of decision-making, characterized by conservatism and strict adherence to rules and procedures. The *Air Ontario Flight Operations Manual,* the company's guiding document for pilots and other operational personnel, did not address hot refuelling. The *Air Ontario Flight Attendant Manual,* however, stated that the Crash and Fire Rescue (CFR) unit was to stand by while hot refuelling was in progress. Although the CFR unit was eventually positioned by the aircraft, the Commissioner concluded that the hot refuelling process began before the fire trucks had arrived and positioned by the aircraft.

Inside the terminal, Capt. Morwood discussed by telephone with SOC the payload, passenger load and IFR alternative. He learnt from

SOC in London that weather in Winnipeg was reported as sky obscured with three miles' visibility in fog, fairly usual operational conditions during the Canadian winter which should have caused no major concern. The information available to the Commission, however, indicated that Capt. Morwood had a heated conversation over the telephone prior to the departure of Flight 1363.

The Cockpit Voice Recorder (CVR) of C-FONF was destroyed in the crash. The CVR would have recorded the conversation in the cockpit during the last instants of Flight 1363. This first-hand information would have provided a better understanding of the reasons why Capt. Morwood made the decision to take off. In the absence of this tool, the Commission felt that the actions of Capt. Morwood inside the terminal in the final moments before he returned to the aircraft were crucial to the human performance aspects of the investigation, and thoroughly inquired into this critical information. The Commission clearly established that the demeanour of Capt. Morwood deteriorated visibly while in the terminal, after his telephone contact with SOC, and that as he briskly walked back to the aircraft, he exhibited undeniable signs of frustration.

Meanwhile, in C-FONF, First Officer Mills established radio contact with the Flight Service Station (FSS) at nearby Kenora, Ontario, and informed the specialist on duty that the visibility in Dryden was about one and a half miles and described the precipitation as 'quite puffy, snow'. He remarked '...it looks like it's going to be a heavy one'. Snow was slowly beginning to accumulate. No one among the survivors or the ground personnel saw either pilots do a walk-around exterior inspection of the aircraft. Although upon his return to the aircraft Capt. Morwood asked the ground handler whether de-icing was available, he did not request de-icing after being told that it was. The ground handler did not further press the issue. When the aircraft was about to leave the terminal platform, snow now falling heavily, its wings were covered in snow to depths varying from one-eighth to one-quarter of an inch. By the time C-FONF entered the runway and began back-tracking towards runway 29, an accumulation of approximately one-quarter to one-half inch of slush covered that portion of the runway.

While taxiing out, FSS advised Flight 1363 of a Cessna 150 in a VFR recreational flight which was due to land at Dryden. The pilot of this aircraft had requested that Flight 1363 hold its departure until he had landed, because of the deteriorating weather. The request was eventually granted and the ground hold further compounded the delay Flight 1363 had been carrying the entire morning. Capt. Morwood is reported to have said over the passenger address (PA)

64

system '...well folks, it just isn't our day...'.

At 12:09 pm CST, one hour and ten minutes behind schedule, Flight 1363 commenced its take-off roll down the slush-covered runway. The combination of the one-half inch deep slush on the ground and the wet snow which had frozen into opaque ice on the forward half of the wings degraded significantly the performance capabilities of the F-28. After a longer than normal take-off roll, C-FONF rotated, lifted off slightly, began to shudder and settled back onto the runway. It rotated a second time, lifting off at the 5700 ft point of the 6000 ft runway. It flew briefly, clearing the end of the runway at approximately 15 ft above the ground. It failed to gain altitude and mushed in a nose-high attitude, commencing to strike trees. C-FONF went down in a wooded area, coming to rest 962 metres from the end of the runway, where it burned (Figure 3.3). It was now up to Mr Justice Moshansky and the professionals supporting the Inquiry to find out why. It would take 22 months to answer this question.

The analysis

In analysing of the crash of Flight 1363, the approach adopted by the Commission to determine the accident trajectory's penetration of the gaps in the system's defensive layers, barriers and safeguards should be kept in mind:

> ...The pilot-in-command made a flawed decision, but that decision was not made in isolation. It was made in the context of an integrated air transportation system that, if it had been functioning properly, should have prevented the decision to take off...there were significant failures, most of them beyond the captain's control, that had an operational

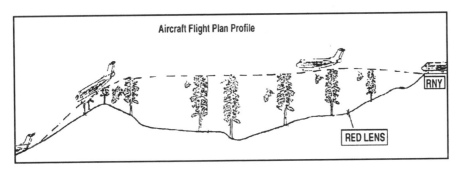

Figure 3.3. Take-off of Flight 1363
(Source: *Final Report of the Commission of Inquiry into the Air Ontario Crash at Dryden, Ontario.*)

65

impact on the events at Dryden...the regulatory, organizational, physical and crew components must be examined to determine how each may have influenced the captain's decision. (Moshansky, 1992, p 1102)

The analysis is worked backwards; from the accident and its surrounding local events to the latent failures and flawed organizational processes which fostered it. The analysis looks into four major areas or blocks: failed defences, unsafe acts, error-producing conditions and latent organizational failures. The conclusions of the analysis are then reflected in the causal pathways model discussed in Chapter 1, to illustrate the active and latent failure pathways which led to the destruction of Flight 1363.

Failed defences

Numerous defences, normal components of the aviation system, should have contained the triggering events on March 10, 1989, and prevented the flight crew from making the decision to take off without de-icing the F-28 wings. Failed defences must never be considered in isolation but within the overall context, since they are symptoms of deficiencies in organizational processes. The most conspicuous defences which failed at Dryden include the following:

- *The unserviceable APU.* The Commission disclosed that Air Ontario personnel often deferred maintenance 'snags' in an unauthorized manner, and subsequently flew the aircraft without fixing them. This led, on a number of occasions, to operate C-FONF without a valid certificate of airworthiness. On the morning of March 10, 1989, the logbook of C-FONF contained six deferred snags, including the unserviceable APU (it would not fire test). Under normal circumstances, operation with an unserviceable APU may be nothing more than an annoyance. Under the operational conditions existing in Dryden on March 10, 1989, it presented a serious dilemma to the flight crew: the lack of an external starting power unit, the hot refuelling, the company policy prohibiting de-icing with an engine running and the problems of shutting down both engines undoubtedly influenced Capt. Morwood's decision not to de-ice at Dryden.
- *Minimum Equipment List (MEL).* C-FONF was dispatched under a provision included in the MEL regarding the failure of the APU fire *extinguishing* system when the actual defect was in the fire *detection* system. Had the cause of the APU defect been correctly listed (or 'MELed'), its use under limited circumstances would

have been permitted. However, the approved MEL was misunderstood and misused by a number of experienced personnel, including a management pilot and a senior maintenance manager who erroneously 'MELed' the defect. This error denied the pilots of Flight 1363 the eventual use of the APU. The Commission determined that the MEL approval process and the use of the MEL itself during daily operations was replete with confusions, oversights and flaws.

- *Information on take-off from contaminated runways.* The information in the Fokker F-28 Flight Handbook regarding take-off performance is based, in accordance with existing certification criteria, on acceleration and stopping taking place on hard, dry and smooth runway surfaces. Accelerate-go and accelerate-stop distances are adversely affected by standing water, slush or snow. At least one-quarter to one-half inch of slush covered the runway at Dryden at the time C-FONF started its take-off roll. The F-28 Flight Manual contains guidance information for operation from contaminated surfaces. So do the USAir and Piedmont F-28 Operations Manuals, both used by Air Ontario pilots depending on where they had received their training. These last two documents provide more restrictive information than the F-28 Handbook, which is in addition time-consuming, difficult and impractical to use within the confined spaces of a flight deck. Had the crew of Flight 1363 used the F-28 Handbook, the take-off at Dryden on March 10, 1989 would have been permitted. Had they used the USAir or Piedmont Manuals, it would not, because the take-off weight would have had to be reduced to 54,300 lbs from its actual weight of 64,440 lbs. But the crux of this matter is that the draft F-28 Operations Manual submitted by Air Ontario to Transport Canada (the regulatory authority) did not include contaminated runway correction charts. Furthermore, Air Ontario had no established policy as to which correction charts (whether those on the Fokker Handbook, the USAir or the Piedmont Manuals) to use; nor was there consensus among the Air Ontario F-28 pilots on the subject.

- *Information on the cold soaking phenomenon.* Cold soaking means that an object has been in a cold temperature long enough for its temperature to drop to, or near to, the ambient temperature. Such is the case of the wing of an aircraft at high altitude and the fuel contained in its tanks. Upon landing, the skin of the aircraft will warm quickly, but not the fuel, which will warm more slowly. The cold soaked fuel touching the wing surfaces will cause the moisture in the air to frost. Rain or wet snow will then freeze to

67

the upper surface of the wing, resulting in an irregular ice surface. On March 10, 1989, the addition of relatively warmer fuel at Dryden raised the level of cold soaked fuel to touch the upper surface of the wing and set the cold soaking phenomenon to work. The Inquiry disclosed that, notwithstanding an insufficient awareness among pilots about the effects of cold soaking of fuel, neither Air Ontario nor Transport Canada had provided pilots with sufficient information regarding this phenomenon and its effect in precipitation and frost adhering to the wing surfaces.

- *Flight following and support.* Weather was a major operational concern for the crew of Flight 1363. Destination and usual alternatives were close to or below operational limits. In addition, because of his limited experience in the F-28, Capt. Morwood had to observe higher take-off and landing minima than those established by Transport Canada in the Canada Air Pilot (CAP), thereby further complicating the selection of alternatives. This in turn had an effect on fuel, passenger and payload. Dispatch – in this case Air Ontario Systems Operations Control in London – can become an invaluable ally to the pilots in these situations by virtue of a knowledge of the 'big picture' which flight crews cannot usually have. On March 10, 1989, Air Ontario SOC did little to contribute to the crew's operational decisions. On the contrary, the flight release available to Capt. Morwood at Winnipeg, prepared by an Air Ontario SOC untrained dispatcher and not familiar with the operational characteristics of the F-28, contained numerous errors which undoubtedly complicated the flight crew's tasks.

- *Reticence of cabin crew and passengers.* The rapid accumulation of snow over the wings during the stop at the terminal, and more notably during the taxi out, was conspicuously noticeable to both flight attendants and to a number of passengers, including two off-duty airline captains deadheading to Winnipeg. In fact, one of these captains later testified to the Commission regarding his incredulity when he realized that Flight 1363 would attempt to take off with such an amount of snow covering the wings. Neither the flight attendants nor the passengers made any attempt to transmit this information to the flight crew. Indeed, Air Ontario cabin crews had previously been discouraged from bringing operational matters to the attention of the flight crew. In the case of the deadheading captains, the Commissioner concluded that professional courtesy probably restrained them – notwithstanding obvious personal risks – from expressing what

68

was covert judgement of value on Capt. Morwood's decision.

- *Crew coordination.* It is a fair statement that the crew members of Flight 1363 did not perform as an intact unit but rather as a collection of individuals. This applies not only to the relationships between the two pilots, but also to the interaction of the flight and the cabin crew. Crew Resource Management (CRM) training, aimed at developing skills for the optimum utilization of the resources available to the captain as well as to other crew members, was not provided by Air Ontario to its crews at the time of the accident. The evidence from the Inquiry suggests that, in addition to the cabin crew, there was at least another 'unhappy camper' on Flight 1363: First Officer Mills. The Report suggests that Keith Mills was concerned about Capt. Morwood's decision to take off, and that he tried to the best of his abilities and within existing etiquette to assert himself and to transmit his concerns. He obviously failed. Had he and the other crew members of Flight 1363 had the benefit of formal CRM training, the coordination among them might have followed different avenues, and the outcome of the situation might have been different.

There is an interesting postscript to this account of failed defences. Even if Flight 1363 had been de-iced on March 10, 1989, the end result might not have been different. According to Canadian practices, the de-icing fluid used would have been Type I fluid, which provides a protection time of 15 minutes under the best conditions. In the conditions at Dryden, this protection could have been reduced to as little as one minute. Type II fluid, which is thickened, provides superior protection. Type II fluid is widely used in Europe and it is considered to be largely responsible for the impressive European record as far as ground icing accidents is concerned. But because it is around 10% more expensive than Type I, its use was not widespread throughout the Canadian aviation community.

Unsafe acts

There is no denying that the flight crew of Air Ontario 1363 failed in its role as last line of defence against system deficiencies. However, not all unsafe acts surrounding the crash of Flight 1363 were committed by the flight crew. Other operational personnel, including ground handlers, dispatchers and even the cabin crew, contributed by their actions or inactions to deny Capt. Morwood and First Officer Mills feedback which could eventually have contained the consequences of their own unsafe acts. Arguably, many unsafe acts were none other than the behaviours fostered by the system, and therefore the behaviours different personnel perceived the system expected from them:

- *No ground de-icing by the flight crew.* The failure to de-ice was clearly the most obvious unsafe act committed by the flight crew of Flight 1363. That Capt. Morwood was of a conservative nature and conscious about de-icing is a matter of record. Earlier on the very day of the accident, for example, he had de-iced C-FONF before departing Winnipeg because of a layer of frost over its wings. He had walked to the terminal in his shirtsleeves; it is impossible that he was not aware of the weather conditions. Therefore, why he did not de-ice at Dryden demands other than a simplistic explanation such as 'with his experience, he should have known better'. Following the line of argument given in Chapter 1, Capt. Morwood did not choose to make a bad decision. His error on March 10, 1989, must be understood within the context and the constraints in which it was made.
- *No walk-around by the flight crew.* Neither flight crew member performed an exterior walk-around inspection. It remains, however, a matter of argument whether such walk-around would have accomplished something, given the inaccurate and incomplete knowledge regarding wing contamination existing among Air Ontario's crews at the time of the accident.
- *Dispatch with unserviceable APU by SOC.* Had he followed the operational restrictions contained in a company's memorandum issued by the director of maintenance, the Air Ontario SOC dispatcher should have advised the pilots of Flight 1363 to overfly Dryden on the day in question, because of the potential necessity for de-icing with an unserviceable APU at a station without ground-start facilities.
- *Inaccurate flight release from SOC.* The flight release provided to Flight 1363 contained numerous errors, including an erroneous maximum take-off weight from Winnipeg, incorrect fuel figures for the revised alternative (Sault Ste Marie) and an incorrect, greater than allowable payload. Similar errors were found in the flight release for the Thunder Bay to Dryden leg. Inaccuracies in flight releases were usual occurrences, and pilots would telephone SOC to notify such discrepancies. Because Capt. Morwood did not communicate any problem to SOC, the Report concludes that throughout March 10 he relied on erroneous information.

- *Revised forecast not disseminated by SOC.* An amended Dryden terminal weather forecast, as well as the Dryden terminal weather forecast issued at 16:30 GMT (11:30 am EST), both called for freezing rain at Dryden during the timespan of operation of Flight 1363. Both were available to Air Ontario SOC while Flight 1363 was still on the ground at Thunder Bay. Such information, which could have induced Capt. Morwood to overfly Dryden, was never transmitted to the pilots of Flight 1363.
- *Failure to follow-up by the ground handler.* There was no follow-up by the ground handler to Capt. Morwood's inquiry about availability of de-icing, even when evidence suggests that he knew that the wings were covered in snow.
- *The cabin crew failure to communicate.* Both flight attendants were aware of the snow covering the wings, although they never attempted to assert this fact to the pilots. As later discussed, serious flaws in organizational processes underlie this unsafe act, including an industry culture which did not (and to a large extent does not) encourage cabin crew to discuss operational matters with flight crews.

In all unsafe acts it is possible to identify numerous contributory situational and task factors (see Chapter 1), such as poor communications, time pressure, inadequate tools and equipment, poor procedures and instructions, and inadequate training. Personal factors (see Chapter 1), such as preoccupation, distraction, false perceptions, incomplete or inaccurate knowledge, and misperception of hazards, are also readily identifiable. It will be later argued in this chapter, however, that flawed organizational processes and latent organizational failures are the source of most unsafe acts committed by operational personnel.

Error-producing conditions

Numerous error-producing conditions led the pilots of Flight 1363 to make the decision to take off without de-icing the wings and other operational personnel to commit their unsafe acts. These conditions are the by-product of latent organizational failures. The Commission demonstrated that latent organizational failures generate not only error-producing conditions. They also have the potential to create a working environment where violations are inevitable if operational personnel are to accomplish their assigned tasks (see Chapter 1).

- *Ambiguous operating procedures.* These ambiguities not only include flight deck procedures, but also maintenance and dispatch procedures. Ambiguities include incomplete information regarding take-off with contamination on the wings and cold weather operations in general, lack of corporate policy regarding hot refuelling and de-icing, and the informal 'blessing' by management of unapproved procedures carried over from the propeller-driven fleet, including a disregard spread among Convair 580 pilots of the effects of wing contamination. These ambiguities are strictly relevant to the events of March 10. In the larger picture, however, the Commission's overall appraisal of the F-28 operation '...reflected operational procedures which...are not recommended in jet operations.' (Moshansky, 1992, p 615)
- *Lack of standardized operations manuals.* Some F-28 pilots used the Piedmont F-28 Operations Manual while others used the USAir F-28 Pilot's Handbook, since Air Ontario did not have its own F-28 operations manual. Although both manuals are comprehensive and both obviously deal with the same type of aircraft, there were sufficient differences in the operating procedures of these two carriers to create potential problems on the flight deck. The Commissioner concluded:

  > After reviewing the... manuals used, and the testimony of many Air Ontario pilots, I have a clear impression that Air Ontario F-28 pilots were often left to learn and to discover for themselves what were the best operational flight procedures for the F-28... an additional and unnecessary burden on the pilots. (Moshansky, 1992, p 568)

- *Training deficiencies.* Aircraft wing contamination, the cold-soaking phenomenon and runway contamination are subjects in which the Commission verified diverging depths of awareness and understanding among Air Ontario pilots. Lack of CRM training is another shortcoming identified by the Report, although '...CRM or equivalent training cannot alleviate operational problems associated with lack of management stability and consistent direction' (p 620). Deficiencies in cabin attendant training, ground handling training and aircraft refuelling training are also discussed.
- *Pairing of inexperienced crew members.* Although both pilots of Flight 1363 had considerable experience, they were 'newcomers' to the F-28. Capt. Morwood had only 62 hrs in the type, and had received his line check on January 25, 1989, after 27.5 hrs of line

indoctrination. First Officer Mills had 66 hrs, having accumulated 29.5 hours of line indoctrination before receiving his line check on February 17, 1989. Although both were legally certificated to operate the F-28, evidence from both accident investigations and research alerts about the dangers in pairing crew members 'new' to the type.

- *Crew frustrations.* March 10 had not been a good day for the crew of Flight 1363. The unserviceable APU and other deferred maintenance items, the confusion over defuelling versus deplaning passengers at Thunder Bay, an inexperienced SOC dispatcher, the absence of ground support facilities, concern over passenger connections and the ground hold for the Cessna 150 are some of the local conditions which fostered crew frustrations. The Report leaves no room for doubt that Capt. Morwood was exhibiting distinct symptoms of stress when he landed in Dryden on the return trip. Stress degrades the ability of humans to process information.

- *Corporate merger and corporate cultures.* Air Ontario is the product of a merger between Austin Airways Limited – a northern or 'bush' operation – and Air Ontario Limited – a Great Lakes, scheduled service operation. Austin Airways was the 'winning party' or buyer; Air Ontario Ltd the 'loser' or acquired company. The two companies were different in almost every aspect: their fleets, their operating environments, their employee groups and their management styles. The harsher demands of flying the Canadian North are qualitatively different from those of flying the 'friendlier' south, demands which were reflected in the experiences of each pilot group. Furthermore, in the non-unionized, northern environment employee responsibilities were rather unstructured, while in the southern, unionized environment employee tasks were clearly delineated. It was not a happy marriage. The two very different corporate cultures were incompatible; yet their effects were enduring and difficult to change, as Chapter 1 explains. Among other conflicts, the negotiations to merge the two pilot groups under the representation of the Canadian Airline Pilots Association (CALPA) ended in a prolonged strike, between March and May 1988. As with any corporate rationalization, resources were greatly taxed. The efficiency with which the various organizational processes were managed is discussed later in this chapter. Of immediate relevance to the events of March 10 is the fact that while Capt. Morwood came from Air Ontario Limited, First Officer Mills came from Austin Airways. The Report includes the contention that the working relationship between the pilots over the previous two days had probably not been cordial, with the subsequent impact on crew coordination.

Latent organizational failures

Three distinct groups of high level management 'contributed' in harbouring the latent organizational failures which eventually led to the crash of Flight 1363: the operator itself, Air Ontario; the regulatory agency, Transport Canada, and the parent company, Air Canada.

Air Ontario Corporate reorganizations generate anxieties among employee groups. In this case, there is evidence of high management turnover, low employee morale and poor job performance, all with potential effects on flight safety. The period following the merger was turbulent. The basic issue examined by the Commission was '...whether Air Ontario management was able to support the flight safety imperative during this period of distraction' (p 407).

In the two years prior to the accident, there had been significant changes in the management of flight operations. There was instability within the flight operations organization, and individuals who had been expected to play a major role in the introduction and management of the F-28 programme had left the company. The Report reveals a situation where effective coordination of efforts had been essential, and which had instead been characterized by a concerning lack of it as well as of effective management: '...deficiencies in (F-28) project coordination were significant to the crash of flight 1363' (p 425).

- *Management turnover and selection.* There were changes in two critical areas in operational management during the period from June 1987 until March 10, 1989: Vice President of Flight Operations and Director of Flight Operations. The instability and problems of supervision created by the lack of management continuity were an obstacle to the implementation of the changes required by the introduction of a new aircraft type. The Report introduces evidence that the President and Chief Executive Officer of the company would personally select all senior management personnel, not always based on the merit principle, but rather in what the Report describes as '...the entrepreneurial management style of a man who has built his company from a small family business' (p 429). Some of the appointments, such as those of the president's close relatives to key managerial positions, were '...the subject of considerable discussion at the Air Ontario committee meeting(s)' (p 431). The outcome of this process was that the operational management of Air Ontario was dominated by individuals whose experience had been mostly in charter operations in the northern, regulated environment, while the new company operated in the southern, deregulated environment as a scheduled carrier. Air Ontario managers were thus confronted by demands for which their experience may not have been adequate.

- *Operational control.* Canadian legislation grants operational control the functions of flight dispatch and flight following, including the authority to initiate, continue, divert or terminate a flight. Operational control provides a crucial support to flight crews by providing updated information to enable them to make safe and efficient decisions. Such control is indeed intended to prevent circumstances like those presented to Capt. Morwood at Dryden. However, '...it was stated by all of the operational control personnel who testified that the training and qualification of the Air Ontario dispatchers was inadequate' (p 728). The Inquiry revealed that when weather was poor, when aircraft had unserviceabilities or under other irregular circumstances – situations in which operational control is an asset – SOC performance usually deteriorated. The Commission concluded that this was a consequence '...of poor planning and organization within SOC, a lack of training and qualification of Air Ontario SOC personnel and the failure of SOC personnel to appreciate the importance of their function' (pp 731–732).

- *The F-28 programme.* The introduction of the F-28 was the first exposure of Air Ontario's management to the operation of a transport category jet aircraft in commercial scheduled service. The management problems discussed revealed themselves in the various flaws and safety shortcomings within the F-28 operation described elsewhere in this chapter, and which can be grouped into two general areas: lack of standard operating procedures, manuals and documentation for the F-28; and inconsistencies and deficiencies in training the F-28 flight crews, cabin crews and ground support personnel.

- *Programme management.* The F-28 project manager had the responsibility to ensure that the implementation and operation of the F-28 programme was properly monitored and supervised. The appointed manager – a relative of the president of the company – lacked experience in the F-28. On the other hand, he

75

was overburdened beyond reason, since he also had responsibilities as F-28 Chief Pilot, F-28 training pilot, F-28 company check pilot[2], Convair 580 Chief Pilot and F-28 line pilot. These combined responsibilities led to an ineffective management of the programme, allowing deterioration of the operational standards below acceptable levels. The Reports describes the project manager as 'a well-intentioned individual' (p 815). It nevertheless must always be remembered that the best intentions, if failed to be carried into deed, are as good as no intentions as all.

- *Maintenance management.* Senior maintenance management was stable during the period June 1987 to March 10, 1989. Nevertheless, numerous maintenance problems were evident in the F-28 operation, including lack of familiarity with the aircraft and a purchase decision which did not include an adequate supply of spare parts. This, combined with the enthusiasm and subsequent organizational overcommitment of the F-28, pressured maintenance personnel and pilots alike to defer and carry maintenance snags for long periods of time.

- *Safety management.* Although mission statements included flight safety as part of Air Ontario's objectives, compelling evidence presented in the Report suggests a rather haphazard approach, *à la* 'safety is everybody's responsibility' (if employees do their jobs correctly, then safety will be optimized). Such a simplistic view – which can be considered nothing more than blowing smoke – denies the technological and sociological realities of contemporary aviation discussed in Chapter 1; realities which impose the imperative of professional safety management. Air Ontario's Flight Safety Officer (FSO) had resigned in late 1987 because of lack of management support, including the lack of access to the Chief Executive Officer. He was not replaced, and the position remained vacant until February 1989, although Air Ontario suffered a fatal accident which involved a DC-3 in November 1988. The Report includes three case studies in the effectiveness of Air Ontario's flight safety organization, which refer to three incidents experienced by Air Ontario. Two of these incidents have common elements with the Dryden crash; the three of them involved the project manager of the F-28 as captain of the incident flights. The Commission endeavoured to

---

[2]A consultant hired to act as carrier check pilot resigned after five weeks of employment. In his letter of resignation, he indicated that much as he would have liked to keep working in the F-28 programme, he could not '...function in my duties as a check pilot when I do not get the support I need'. (p 613).

evaluate how the Air Ontario flight safety programme, or its lack thereof, had an effect on the F-28 operation, and to identify possible links with the crash of Flight 1363. The evaluation led the Commission to conclude

> ...the lack of continuity in the position of a flight safety officer, the lack of adequate support of the FSO position by senior management and the lack of a flight safety organization over the material time span was a managerial omission... The management assigned a low priority to the importance of filling the vacant position of FSO... This period of instability carried over into the introduction of the F-28 programme had an impact on flight safety. (Moshansky, 1992, pp 785–787)

The Commissioner's assessment of the corporate state of affairs and their influence on the events of March 10 1989 leaves no room for doubt:

> The Dryden scenario, in my view, was reasonably foreseeable (p 531)... Air Ontario was not ready in June 1988 to put the F-28 aircraft into service as public carrier. (Moshansky, 1992, p 815)

Transport Canada   Transport Canada is the Federal agency tasked with the promulgation and enforcement of aviation regulations; the provision of air navigation and airport services; the certification of personnel, carriers, aircraft and aeronautical products; and the surveillance, inspection and audit of the system. It is responsible to the people of Canada for ensuring that aviation is carried out effectively at an acceptable level of safety.

The 1980s had not been a sympathetic decade to Transport Canada. Economic deregulation – a policy of the Federal government – had brought a considerable increase in workload, while at the same time, measures aimed at Federal deficit reduction had led to downsizing of the workforce. This produced a situation where the agency saw its ability for surveillance, inspection and monitoring greatly reduced. Numerous warning flags were raised by different sectors within the Canadian aviation industry, with little or no effect. The decrease in personnel, inadequate training policies and supporting programmes, and mismanagement of human resources led to a state of affairs where Transport Canada was in a very precarious position to discharge its responsibilities in a timely and effective manner.

- *Regulatory audit of Air Ontario.* An audit of Air Ontario by Transport Canada was scheduled for February 1988. While the airworthiness, passenger safety, and dangerous goods portions of the audit were completed as scheduled, the operations part of it was postponed because Air Ontario did not have an approved flight manual in place. The operations portion was re-scheduled for June 1988, and eventually conducted between October and November 1988. Because the audit team leader had no jet experience, the audit did not cover the F-28 programme. The Report consider this '...a serious omission. Had the F-28 been audited, it is reasonable to assume that a number of deficiencies relating to Air Ontario's F-28 operation would have been discovered prior to the Dryden crash' (p 997). Notwithstanding Federal policy to release audit reports within 10 working days of the completion of the audit, Air Ontario was not presented with a report until after the crash, more than five months after the audit had been completed. The Report characterizes the Transport Canada audit of Air Ontario as '...poorly organised, incomplete and ineffective' (p 996).

- *The operating rules.* The Inquiry accumulated evidence that existing regulations applicable to Canadian air carriers were '...deficient, outdated and in need of overhaul and outright replacement' (p 1000). Areas in which deficiencies were identified or in which there existed no regulations at all included flight dispatch requirements, minimum equipment lists, shoulder harnesses for flight attendants, approval of aircraft operating manuals, and qualifications for air carrier managerial personnel. The hearings also disclosed ambiguity of aviation regulations and air navigation orders. In the case of the MEL, none of the witnesses could define with reasonable precision one of its most critical terms: essential airworthiness item. As another case in point, the pilots of C-FONF carried two aircraft operating manuals, different in form and contents, and without amendment service (Capt. Moorwood had the Piedmont manual and F/O Mills had the USAir manual). Neither manual was approved by Transport Canada, since no regulatory requirement existed to this effect. Transport Canada operational staff who testified at the Inquiry were unanimous in their views about the inadequacy of existing regulations and '...the chronic inaction on the part of Transport Canada senior management in many areas of urgent concern...' (p 1006).
- *Safety management.* The Inquiry revealed that because of resource constraints, an inadequate regulatory framework and organizational deficiencies, Transport Canada was not ideally able to ensure an efficient and uniform level of safety.

Deficiencies uncovered included distinctly separated lines of reporting to the top of the organization and the apparent inability of different internal groups to work together in identifying and addressing safety issues. The Commission also expressed concern that Transport Canada was '...spending too much energy on minor violations that were of little safety consequence, while not enough effort was being put into overall education and safety promotion' (p 1043).

That Transport Canada

- did not provide clear guidance for carriers and crews regarding the need for de-icing;
- did not enforce the provision of performance data on contaminated runways;
- did not closely monitor Air Ontario for regulatory compliance following the merger and during the initiation of the jet service;
- did not require licensing or effective training of flight dispatchers;
- did not provide clear requirements for the qualification of candidates to management positions, including director of flight operations, chief pilot and company check pilot;
- did not develop a policy for the training and operational priorities of air carrier inspectors;
- delayed the audit of Air Ontario and did not include the F-28 programme in it;
- followed an excessively complex MEL approval process; and
- did not have a clear definition of what constitutes an essential airworthiness item

are oversights and flaws that nurtured the trajectory of opportunity, and which combined with local triggers at Dryden on March 10, 1989, to break the system defences, safeguards and barriers and permit the accident.

The Commission concluded:

...Air Ontario, as a commercial air carrier, was not operating in a vacuum. Transport Canada, as the regulator, had a duty to prevent the serious operational deficiencies in the F-28 programme...Had the regulator been more diligent in scrutinizing the F-28 implementation at Air Ontario, many of the operational deficiencies that had a bearing on the crash of flight 1363 could have been avoided. (Moshansky, 1992, p 438)

Air Canada   The controlling interest of Air Ontario was owned by Air Canada. Air Ontario was marketed as part of Air Canada's network, and a public perception of an integrated company had been fostered. Air Canada dedicated a significant effort to present a close integration in the marketing functions. These marketing efforts had been rewarded by a measure of success; many of the passengers of Flight 1363 believed that they were in fact flying with Air Canada. In specific relation to the crash of Flight 1363, the Commission raised the issue of the lack of application of Air Canada's expertise in scheduled jet operations to the Air Ontario F-28 programme. The evidence in the Report reveals that '...these initiatives were not in any way directed towards verifying and monitoring the operational procedures and flight safety standards of its new subsidiary. On the contrary, Air Canada deliberately maintained its corporate distance from the operational end of Air Ontario' (p 818).

The regulatory standards defined by Transport Canada – and by any other civil aviation administration – represent minimum standards, referred to in the Report as 'the threshold level of operational safety'. The evidence demonstrates that Air Canada operated at a greater level of safety than that required by Transport Canada. The evidence also demonstrates that Air Canada management '...while imposing on Air Ontario its own high marketing standards, required Air Ontario only to comply with Transport Canada's threshold operational safety standards' (p 835).

The Report discusses Air Canada's lack of support to Air Ontario during the introduction of the jet service, and compares standards in specific areas such as operational policies for dispatch with an unserviceable APU; minimum equipment lists; manuals; aircraft defects; hot refuelling policies; de-icing policies; operational control and flight planning; and dispatcher training. The evidence demonstrates double safety standards between Air Canada and Air Ontario.

The Report also reviews Air Canada's flight safety organization and its involvement with Air Ontario. In spite of the fact that Air Canada had significant experience in introducing jet service (on several types), this experience was not made available to Air Ontario when it introduced the F-28 service. The assistance Air Canada planned to provide its connector was limited to the provision of information relating to flight safety and playback facilities for flight data recorders. In practice, this intention was further reduced to a post-accident response seminar in 1985 and another in May 1989, three months after Dryden. The Report also describes with detail Air Canada's flight safety organization, leaving no room for doubt about its importance and the corporate commitment which supported it. The double standards again become obvious when reviewing Air Ontario's flight safety organization. The director of flight safety for Air Canada testified that he was under the impression that Air Ontario had a flight safety officer. It did not. He also assumed that computer recording and trend analysis was being carried out by Air Ontario. It was not. When asked about the degree of integration between the flight safety organizations of parent and feeder, he conceded that there was none. Lastly, the representative of Air Canada on the Board of Directors of Air Ontario appeared to be unaware that, for more than one year and during the crucial timeframe of the F-28 introduction, there was no flight safety officer or flight safety organization in Air Ontario.

The Report summarizes: '...The evidence also demonstrates that Air Canada had little involvement in the flight safety aspects of its subsidiary, Air Ontario...Furthermore, Air Canada did not impress upon Air Ontario its own more developed flight safety ethic' (p 784).

In conclusion:

> The corporate mission statements of Air Canada and Air Ontario both contain words to the effect of the primacy of safety considerations. The evidence disclosed that other corporate concerns, important in their own right, were allowed to intervene and subordinate safety. *The difference between the attention and resources expended by Air Canada and Air Ontario on marketing, as compared with safety of operations, must, when held up to their respective mission statements, be described as inadequate and short-sighted.* (Emphasis added) (Moshansky, 1992, p 853)

The causal pathway

When the the previous analysis is reflected in the causal pathway model, the visual presentation shown in Figure 3.4 emerges. That latent failures become markers of the 'state of health' of the system (see Chapter 1) appears obvious in Figure 3.4. If, as Chapter 1 suggests, we could somehow 'freeze' the system and remove the left-hand side of the model below the box labelled 'local working conditions', that is, if the active failures were removed and the crash of Flight 1363 did not take place, the flawed organizational processes and the latent failures would still remain. Dryden would take place elsewhere, given the degree of sickness, the numerous pathogens hidden in the system. On the contrary, if we could remove the right-hand side of the model, the left-hand side would also disappear. The implications for prevention endeavours should be clear.

81

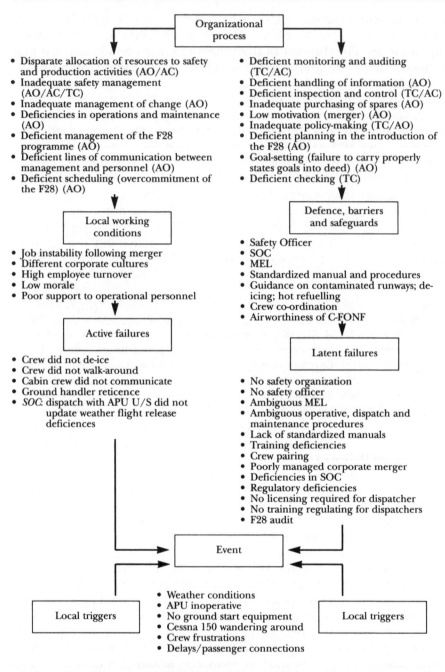

Figure 3.4.   The causal pathway (AO: Air Ontario, AC: Air Canada; TC: Transport Canada)

82

Conclusions

When the moment arrived to close the file, the evidence obtained and discussed over 20 gruelling months led Mr Justice Moshansky to conclude:

> Captain Morwood, as the pilot-in-command, must bear responsibility for the decision to land and take off in Dryden on the day in question. However, it is equally clear that the air transportation system failed him by allowing him to be placed in a situation where he did not have all the necessary tools that should have supported him in making the proper decision. (p 1131)

This statement holds the key to securing and advancing safety and effectiveness in modern, complex sociotechnical systems. Morwood, Mills, the cabin crew, SOC dispatchers, the ground handler and other personnel involved in the operational events surrounding Flight 1363 failed in their role as last line of defence, and thereby precipitated the accident. For this, they must be held accountable. If we are looking for scapegoats, we need go no further. But if what we seek is to avoid future tragedies like Dryden rather than sweeping the dust under the rug, we must examine the organizational processes which generate gaps in the system defences and induce properly qualified, healthy and well-intentioned individuals to make such damaging mistakes. The message from the Report is two-fold. On the one hand, there should be no doubt: there is still no substitute for a properly trained, professional flight crew. They are the goalkeepers of aviation safety. But on the other hand, no matter how hard they try, no matter how professional they might be, no matter their care and concern, humans can never outperform the system which bounds and constrains them. If in contest, system flaws will sooner or later inevitably defeat individual human performance.

A postscript

The *Final Report of the Commission of Inquiry into the Air Ontario Crash at Dryden, Ontario,* and the process of the Inquiry itself, were the subject of a domestic controversy in Canada. The Report provoked mixed feelings within the Canadian aviation community. This came as no surprise to alert observers, since such is usually the case whenever change is about to take place. Make no mistake: *The Canadian aviation system failure on March 10, 1989, was to a large extent fostered by deficiencies* embedded in its design, and its performance on that particular day was somehow to be expected by virtue of its underlying flaws. The Report acknowledged these facts, and thus became the starting point for implementing change, not only within the Canadian aviation system, but largely in established aviation accident investigation and accident prevention practices elsewhere.

In the final analysis, the controversy is inconsequential. It is, quite understandably, part of the denial processes into which many of those who unwillingly or unknowingly fostered the accident pathway embarked the very moment the Report was made public, forgetting the simple fact that no matter how intensely we might try to deny reality, reality prevails. The reality of March 1989 was that – with or without Dryden, with or without Commissioner Moshansky's Report – there existed an aviation system with numerous flaws which, for many different reasons, had remained undetected for some time. These flaws eventually led to the crash of Flight 1363 and to the loss of 24 lives. The Report simply brought such flaws into the open, so that they could be properly addressed.

In closing, a caveat is felt necessary: this chapter may convey to readers who are not familiar with the Canadian aviation system and its organizations the notion of something akin to a multiple-headed monster lurking in the darkness for its prey. Such a notion could not be more removed from truth. The Canadian aviation system, notwithstanding all the flaws Commissioner Moshansky and his team uncovered, is one of the safest in the international aviation community. Furthermore, Transport Canada is one of the most pro-active, safety-minded civil aviation administrations of the 182 Member States of ICAO, the International Civil Aviation Organization. Both facts only strengthen this book's argument about the inherent weaknesses of large systems where, like in aviation, people and technology interact actively in pursuing production activities. They also suggest the need for organizational approaches to ensure contemporary aviation safety and effectiveness.

Six years after Dryden, many have learned its lessons. As Chapter 1 suggests, Dryden is the dawning of an age. Most notably Canada and Australia, as the true pioneers of systemic approaches to accident investigation (see Chapter 5); and the United States, Finland, Sweden and France following the ICAO's adoption of the Reason model as a prevention tool, are among the countries whose accident investigation agencies have embraced the organizational approach to accident investigation. In so doing, they are leading the international aviation community towards a safer and more efficient aviation system.

Acknowledgements

We would like to acknowledge Peter G. Harle for his careful, methodical and insightful review of the draft of this chapter and his highly focused feedback. We would also like to acknowledge Stephen Corrie for his review and comments on an early draft.

Reference

Moshansky, Mr. Justice (1992) *Commission of Inquiry into the Air Ontario crash at Dryden, Ontario.* Final Report, Ottawa, Canada.

4 The BAC1-11 windscreen accident

Introduction

Whenever we wish someone 'good luck' we acknowledge that no matter how hard we try to cover all the angles, our lives are subject to random influences, sometimes benign and sometimes not. This case study shows both faces of chance.

In seeking explanations for puzzling events, the human mind '. . . is prone to suppose the existence of more order and regularity in the world than it finds' (Bacon, 1620). One way of simplifying the world is to assume a symmetry of magnitude between causes and consequences. In the face of horrific man-made catastrophes like Bhopal, Chernobyl, Seveso and the *Herald of Free Enterprise*, it seems natural to look for some equally monstrous act of irresponsibility or incompetence as the primary cause. But a detailed study of major accidents in complex systems rarely uncovers any gross failings or obvious villains. What we usually find is the chance and largely unforeseeable concatenation of many different causal factors, none of them sufficient nor even especially remarkable by themselves, but each necessary to bring about the final outcome.

Chance does not make moral judgements. Bad luck can afflict those who deserve better. This was true, we believe, of the British Airways maintenance engineers at Birmingham whose system of working and a sequence of erroneous actions resulted in the near-catastrophic blowout of the left windscreen. Good luck and prompt action by the crew played a large part in saving the lives of the Captain, who was partially sucked out of the window, and of his passengers and crew.

86

One of the paradoxical features of this accident was that many of the contributing factors were rooted in what, under other circumstances, would be regarded as valuable strengths. Some of these features are listed below:

- A Shift Maintenance Manager (SMM) who did not like the look of the bolts he was removing from the old window and so took the trouble to replace them all with new ones, despite his problems in tracking them down. If it can be a fault, the SMM was highly conscientious and liked getting his hands dirty, even though the bulk of his duties were supervisory and administrative. He was known for working long hours of overtime and for his high level of commitment to the work.
- A close-knit establishment of maintenance engineers at Birmingham serviced 13 elderly BAC1-11s, comprising the local BAC1-11 fleet, and worked on HS 748s, ATPs and a variety of third-party aircraft. Both the workload and the morale were high. A good deal of competitive pride was felt by the five rotating shifts in the amount of work they could get through. Such was this enthusiasm that the local management had repeatedly cautioned the night shifts, in particular, not to attempt more than was prudent in the time available.
- An excellent relationship existed between the flight crews and the maintenance personnel. Both were dedicated to keeping the aircraft safe and flying on schedule. However, since the manuals and the Illustrated Parts Catalogue (IPC) were sometimes out of date, the engineers relied to a certain extent upon professional judgement. In short, the workforce saw their job as responding flexibly to the rapidly changing demands on their skills. But this did not mean getting the aircraft away whatever the cost. They operated on an 'any doubts, then ground it' principle. The work entailed a good deal of 'thinking on their feet', and they enjoyed this aspect of the job.

Outline of the accident

On 10 June, 1990, British Airways (BA) flight 5390, en route for Malaga, was climbing through 17,300 feet on departure from Birmingham when the left windscreen was blown out under the effects of cabin pressure. The Captain was sucked halfway out of the aircraft, but was restrained by cabin crew while the co-pilot made a safe landing at Southampton Airport. The Captain remained half out of the windscreen frame until the aircraft landed. He suffered bone fractures in his right arm and wrist, a fractured left thumb, bruising, frostbite and shock. There were no other serious injuries.

The aircraft was a BAC1-11 Series 528L. It was manufactured in 1977 and had a total of 32,724 airframe hours, with 41 hours until the next check.

The windscreen had been replaced just prior to the flight by the Birmingham maintenance facility. It was later discovered that 84 of the 90 bolts securing the window were smaller than the specified diameter. In outline, the events that led up to this near-disaster were as follows:

- The SMM, on a short-handed night shift, decided to carry out the replacement of the left windscreen himself. He consulted the maintenance manual to refresh his memory, and gained the impression that it was straightforward job.
- With the help of the avionics supervisor, he removed the old window and decided that because some of the bolts were paint smothered or damaged on removal he would replace all of them.
- Taking a removed 7D bolt with him, he searched through the drawers of the supervised carousel and accurately matched (by sight and touch) the removed bolt with a drawer's contents. But there were only 4–5 bolts remaining in the drawer, and he needed over 80.
- He drew the storeman's attention to the fact that the drawer contents were below the required stock minimum of 50. The storeman told him (correctly) that the job required 8Ds, but did not press the matter. (There were also insufficient 8Ds in the carousel, though they were not searched for.)
- The SMM decided that since 7Ds had come out of the old windscreen, he would replace them with the same bolts.
- The SMM drove some two miles back to the International Pier, where his office was located. His intention was to collect the requisite number of 7Ds, plus six longer bolts for attaching the windscreen to the outer corner fairing strips.
- Holding his removed 7D bolt between finger and thumb for comparison, he checked it against the contents of a nearly full drawer in the unsupervised carousel. He judged the old and the new bolts to be a match. Unfortunately, the new bolts were 8Cs, longer but thinner bolts than his target 7Ds.
- The SMM returned to the hangar and replaced the 60 lb new windscreen single-handed, using the incorrect bolts. Since they went into elliptical nuts, he did not detect any problems when torquing the bolts home. Nor did he notice the large amount of

countersink visible after the bolts had been secured.

Causal analysis

In keeping with the practice established earlier in this book, the causal analysis begins with a consideration of the absent or failed defences. It then goes on to discuss the unsafe acts committed by the SMM, the local error- and violation-enforcing conditions, and the organizational and managerial latent factors.

Absent and failed defences

There is one key issue here: how was it possible for the quality lapses in the installation of the new windscreen to remain undetected before it blew out at 17,000 feet? Several factors conspired to produce this near-catastrophic path through the elaborate defences surrounding aircraft maintenance in British Airways and other major airlines:

- The British Civil Airworthiness Requirements relating to the BAC1-11 Model 500 call for duplicate inspections after certain critical maintenance operations. In 1985, the Civil Aviation Authority (the regulator) introduced a requirement for duplicate inspections of 'Vital Points'. These are defined as 'any point on an aircraft at which a single mal-assembly could lead to catastrophe, i.e., result in loss of the aircraft and/or fatalities' (AAIB, 1992). However, this did not cover the windscreen replacement or other multiple fastenings. But, as the accident investigators pointed out, the windscreen was probably unique: first, in being able to accommodate the incorrect installation undetected and, second, in exposing these failures so dramatically the moment it was required to withstand the pressure differential. 'Had it been any other item, the selection of the wrong bolts may have been unmistakably apparent during the fitting process, or the subsequent failure may not have been so obvious or traumatic' (AAIB, 1992).
- The windscreen was designed to be secured with countersunk head bolts to British Standard (BS) A211-8D. The British Standard specifies that the BS number and the bolt part number should be shown on the label of the parcel of bolts, rather than upon the bolts themselves. The 8C bolts were found loose in the carousel drawer.
- There were no CAA or Maintenance Manual requirements for a

pressure check to be carried out after a window installation.

- The licences held by the SMM, his supervisory role, the policy of self-certification, and the company authorizations granted by the CAA meant that he was able to release the aircraft without duplicate checks being made upon his work. It only required a short statement – 'Windscreen replaced. Air Safety Report actioned. Functional check (of windscreen heating system) satisfactory' – on a preprinted Release to Service Certificate and a sign-off by the SMM himself to clear the necessary paperwork. The SMM was the only person on the night shift whose work was not subject to the review of a maintenance manager.

No system of defences is ever perfect. Each one of these absent defences was a tiny chink rather than a yawning gap. Each defence alone had stood the test of time and practice. It was only the highly unlikely coincidence of all of these separate chinks on the night shift in question that created a pathway of accident opportunity.

In its investigation report, the Air Accident Investigation Branch identified other weaknesses in the barriers and safeguards surrounding aircraft maintenance at Birmingham. These stem from their conviction that the unsafe acts involved in the windscreen fitment were not 'one-offs' but were symptomatic of a more widespread slippage of procedural compliance and quality standards:

- An inspection by the CAA's Flight Operations Inspectorate had taken place approximately one year earlier. It had lasted half a day, but had not detected any serious engineering problems. The AAIB believe that this inspection was too superficial to detect the presumed slippage of standards.
- A physical British Airways Quality Audit, required under their Quality Monitoring Procedures, had occurred two years earlier (paperwork audits occur twice yearly). The opinion of the audit team, following a two-day visit, was that the engineering facility was to a high standard. Seven observations were raised relating to minor, non-aircraft, matters. The inference to be drawn from the AAIB report was that this audit too was inadequate to reveal the supposed system failings.
- The Quality Monitoring Procedure had been introduced by British Airways in 1987. It had three main components: a local exposition that lists functions and locations of work; continuous monitoring of defects through the Quality Monitoring Deficiency Report (QMDR) system; and product sampling, carried out at set periods, with reports sent to the Chief Engineer

for Quality. The AAIB argued that these procedures should have identified some of the situational factors contributing to this accident: the poor labelling and segregation of bolts in the uncontrolled carousel under the International Pier, the inadequacies of the work platform (safety raiser) and the inappropriate torquing tools.

- The Area Maintenance Manager, the AAIB claimed, failed to monitor directly the work standards of the engineers at Birmingham. The Station Maintenance Manager failed to monitor directly the standards used by individual Shift Maintenance Managers. Both of these assertions are disputed by British Airways senior management.

This is not the place to discuss whether or not these last items represented valid or truly influential defensive failures. We will consider this more fully at the end of the chapter, in relation to some broader issues associated with accident investigation.

Unsafe acts

There is no ambiguity regarding the unsafe, or at least deviant, acts contributing to the misinstallation of the windscreen. All parties to the investigation, including the SMM, agreed upon the following:

- The failure to consult the IPC to establish the correct bolt size.
- The failure to heed the storeman's comments about the 8Ds.
- The inappropriate methods used to select the bolts, i.e. by look and touch comparisons.
- The misdiscrimination of the bolts at the unsupervised carousel.
- Setting a torque of 20 lbf. in instead of the specified 15 lbf. in.
- Failing to detect the attachment of the wrong bolts, either through the bolt head countersink discrepancy or through torque 'feel' cues.
- Failing to recognize the significance of his actions when, on the following day, he fitted a windscreen with 8Ds already supplied. The aircraft with the incorrectly installed windscreen had not yet flown.

To this agreed list of active failures should be added another: the SMM's decision to undertake the window fitting at all, and then performing the task largely single-handed. Each of these active failures is considered in detail below.

Failure to consult IPC This was not so much an error as a violation. These two kinds of unsafe act are psychologically quite distinct, although they are sometimes difficult to differentiate from accident reports. Violations are deliberate deviations of planned actions from regulated or required practice. Where there is no malicious intent (as in this case), they are committed for what seem like a good idea at the time. Perhaps the commonest form are *routine violations*: taking short cuts between task-related points.

In his statement, the SMM made it quite clear why he chose to take this short cut: 'I was satisfied that the bolt I removed was the correct bolt and it takes some time to find the correct part numbers from the IPC which would take up time I did not feel was justified in the circumstances of the job in question.' These may not be laudable sentiments, but they are perfectly understandable. After all, at least 78 of the bolts securing the old windscreen were 7Ds and he had correctly established their identity at the carousel. Moreover, they did have three clear threads and had performed the job adequately for the past four years. Nor did they not 'feel short' on removal.

In an interview with the SMM some two years after the accident, he was asked how often in his work was it necessary to rely upon judgement because the manual and its associated IPC were out of date or the information poorly presented. His reply indicated that this was quite frequently the case. He explained that the IPCs for the BAC1-11s inherited from British Caledonian had had a number of problems, and that considerable time had been devoted to putting them right. Again, this does not excuse his failure to consult the IPC, but it does provide some grounds for the violation.

At this point, it is worth considering what might have happened had he consulted the IPC. First, the relevant illustrated section of the IPC showed only a 7D securing the windscreen in question. It is possible that a glance at this picture might simply have confirmed his original identification. But suppose he had read the small print and correctly identified the 8Ds; would this have thwarted the accident sequence? Not necessarily. As it turned out, the 8D section of the supervised carousel, like the 7D section, was nearly empty. It is then possible that he would have gone to the unsupervised carousel, adopted his look and touch method of selection (given the state of the carousel there was hardly any local alternative) and selected the 8Cs just the same, though this would have been less likely had he been matching drawer contents to a new 8D. In their clean state, they are readily discriminated by the bolt head diameters.

Another issue raised in this context is why the SMM did not consult the computer-based stores inventory system, TIME (Total Inventory

Management for Engineering). In theory, this system is such that an item whose part number has been identified can be located down to the drawer containing the stock. In practice, TIME was relatively new and still not wholly trusted. There were often discrepancies between the reality and the stock levels recorded, because people removed items but sometimes failed to book them out through the computer.

Failure to heed storeman's comment Both this and the failure to consult the IPC have the character of *rule-based mistakes*. These are errors that involve the misapplication of a normally good problem-solving rule because of a failure to notice local counter-indications (Reason, 1990, pp. 74–79). In this case, the experientially-based general rule was 'replace like with like' and the counter-indications were the IPC data and the storeman's comment. The reasons for not consulting the IPC have been discussed; the reasons for not acting upon the storeman's remark have a good deal to do with the personalities of the two people involved, and with the nature of the interchange between them.

In an interview with the SMM, he was asked: 'Did you hear and take in the storeman's comment about the 8Ds? If so, why did you not act upon it?' He replied: 'Yes, but it [the storeman's remark] was more of a parting shot than a comment. The storeman is good at his job, but he's inclined to tell us how to do ours. Sometimes he's right and sometimes he's not. He's very sensitive to criticisms about store's procedures. His remark – "Well most people use 8Ds for that job anyway" – was a comeback after I'd criticized the low 7D stock level. Ironically, the 8D drawer was empty as well.'

Inappropriate method of bolt selection In the event, this ill-judged method proved to be the SMM's main undoing. However, it is difficult to see what other method he could have adopted at the unsupervised carousel, given his intention to obtain the bolts immediately in order to finish the job in time, and the state of the unsupervised carousel.

Misdiscrimination of the bolts This is a perceptual slip, a failure to discriminate very similar items in conditions of poor illumination and with imperfect vision. The SMM described what happened when he reached the unsupervised carousel: 'I started looking in the place where I thought the 7Ds would be. I opened the drawers and inspected their contents. I had the removed bolt in my hand and matched it up with a new bolt by placing them side-by-side on my left hand and then putting my right thumb and forefinger top and bottom of the two bolts to check their lengths. The only way I can explain how the mismatch came about was that the old (removed) bolt was black and had a paint

93

ring on it. This paint ring made the diameters of the heads seem much more alike than they actually were.'

The 20 lbf. in. torque setting   The SMM's experience of loose nuts suggested that an extra torque would have been safer. If this actually is an error, then it would be a *considered but mistaken judgement.* It should be stressed that, however unwisely, the SMM was employing his engineering judgement to achieve a safer aircraft.

Failure to detect wrong bolts   This was the error that the SMM himself found hardest to explain. Again, it was a failure of discrimination. His comments were: 'I don't know how I didn't spot the bolt head and countersink discrepancies. I really blame myself for that. There's no real explanation I can offer. Perhaps I was distracted about the time and about the progress of other ongoing jobs.'

Lack of awareness on the next day   In view of the nature of the inherited BAC 1-11 fleet, the SMM did not find such a difference in method of attachment at all surprising, i.e. between what he supposed were 7Ds the day before and the 8Ds he was currently using. He also commented that in the past, during his service in the Royal Air Force, the basic rule was never to use a bolt that was too long, since there was a danger of rupturing the pressure seal. The 7Ds that he thought he had used on the previous day seemed right, particularly as they had three clear threads beyond the nut.

Undertaking the job himself   From the remarks they made during the investigation, it seemed that not all of the other SMMs would have undertaken this task themselves. Here it appears that the SMM's personality played a significant role. If it can be a fault, the SMM was over-zealous about his work. He also liked 'to get stuck in'. And, as he himself conceded, this was probably unwise. In his own words, it probably affected his ability 'to stand back from a job'. If he had taken a more detached view, then he might not have been so worried about finishing the windscreen replacement in time for the wash that had been booked for 06.00 hrs that morning. Saturday was a quiet day. The job could have been postponed. But fresh in his memory were the problems he had experienced with the ramp crew boss when a wash had been cancelled a week earlier. He did not want a repetition. This, and his hands-on involvement, seem to have affected his judgement adversely with regard to the wisdom of taking on the windscreen replacement alone and largely unaided. He was a man who took a lot on to himself, and as events showed, this was sometimes too much.

One final quote from the SMM: 'I was certainly distracted by having so many balls in the air. When you are looking at the time and wondering how the other jobs on the shift are going, your mind can wander away from the job in hand.'

Error-producing conditions and local performance-shaping factors

Errors do not usually emerge out of the blue. An examination of their surrounding conditions usually reveals a number of local factors that promoted their occurrence or impeded their detection. Some of the principal error-enforcing factors at work in this accident sequence are listed below:

- The unsupervised carousel under the International Pier was poorly and inaccurately labelled. Only 163 of the 408 drawers contained solely the contents described on the drawer label. In the event, of course, the SMM did not consult the labels, but matched the bolts by sight and touch.
- This matching process was impaired by two factors: first, the area was poorly lit and the light was behind the SMM as he attempted to match the bolts; second, the SMM was presbyopic and required glasses for close work. He possessed a pair of half-eye reading glasses, but was not wearing them at the time.
- The calibrated dial-indicating torque wrench was not available for use that night. However, the storeman had recently acquired, on his own initiative, a torque limiting screwdriver specifically for the windscreen task. On arrival, though, it was found to be out of calibration date and not cleared for use. The storeman intended to send it to the London standards room for resetting, but in the absence of any other suitable tool, he set the screwdriver to 20 lbf. in. as requested by the SMM, and issued it to him for the windscreen job. The SMM checked the setting twice using both torque checking gauges.
- The SMM used a quarter-inch bi-hexagonal socket to hold the No. 2 Phillips screwdriver onto the speedbrace. But the socket failed to retain the screwdriver bit, so the SMM had to hold it in place with one hand. During the course of installing the windscreen, the bit fell out several times and he had to get down from the safety raiser (the mobile platform giving him access to the job) to retrieve it. To carry out the task, he had to stretch across the nose of the aircraft and his left hand (holding the bit into the socket) obscured his view of the bolt head.
- The safety raiser was incorrectly positioned alongside the

aircraft. This meant that he had to lean outside the safety rail to reach the job. Due to this uncomfortable position and his left hand obscuring the bolt heads, the SMM was not able to observe directly that the bolt thread was slipping in the anchor nut thread instead of the torque limiting screwdriver, allowing its shaft to remain stationary while the handle rotated. The feel obtained from the screwdriver as he ran the bolts home was what he expected. And the position of his hands limited his vision of the discrepant coutersink.

These situational factors had three important consequences: first, they promoted the likelihood of a mismatch at the carousel; second, they made it harder for him to detect this error; third, they contributed to his failure to realize his error during the installation process.

Organizational factors

Organizational factors originate 'upstream' from the various workplaces. They are the consequences of decisions, concerning design, manufacture, management, standards and procedures, strategic goals, regulation and the like, that can create the workplace conditions that may provoke individual errors and violations. Some of these have already been discussed under the heading of 'failed defences', others are listed below:

- The windscreen to be replaced had been fitted with 7Ds. They had proved adequate for the job for some years, but they were not the recommended fastenings. The fact that the SMM used a correctly recognized 7D bolt as the target for his look and touch method of search clearly played an important role in leading him down his error-prone pathway.
- Had the SMM consulted the relevant IPC page (available on a microfiche reader), he would not have found it a model of clarity. It showed a sketch of the pilot's No. 1 windscreen and the adjacent direct-vision (DV) window, but the only bolt actually illustrated was an A211-7D. The components for the window were shown in the text along with several modification states. The bolts were identified there as A211-8Ds. It is quite possible that a glance at this page would merely have confirmed the SMM's belief that the removed 7D was the appropriate bolt. But even if he had read the small print and identified the correct 8Ds, the drawer in the supervised carousel containing 8Ds was also incomplete. Under those circumstances, he had two choices:

either order an adequate supply of 8Ds and delay the job until they arrived, or proceed as he did to search through the unsupervised carousel – with possibly the same result. Given his sense of time pressure (due to a perhaps unnecessary concern about an aircraft wash booked for 06.00 hrs that morning), he would probably have adopted the latter course.

- The TIME (Total Inventory Management for Engineering) system had recently been installed in Birmingham, but was not widely trusted by the workforce, as mentioned earlier.

- It need hardly be said that the accident could have been avoided had the stock levels in the supervised carousel been adequate.

- The existence of the uncontrolled carousel, its inaccurate labelling and the non-procedural way in which the SMM went about obtaining the bolts all indicate a basic problem in the management of aircraft general stores (AGS) at Birmingham.

- The SMM's job was primarily administrative and supervisory. On occasions, however, he also carried out 'hands on' work. This mixture of roles is an uncomfortable one, dividing attention in the case of 'hands on' work between monitoring other engineers and carrying out the task in hand.

- The errors committed by the SMM could have been avoided had the new windscreen come with the correct fastenings attached. Multiple attachment items, like windscreens, could be packaged as a unit that includes the requisite fastenings.

- Although all designs are a compromise, it does appear that having windscreens attached to the outside of the airframe, when the pressure loading comes from within, was revealed in this instance as a latent design weakness, though not one unique to the BAC1-11. However, BA management have suggested that plug fits of this kind may be more susceptible to bird strikes.

- The AAIB identified what they believed to be organizational deficiencies in the reporting and appraisal structure. The Area Maintenance Manager (AMM), they asserted, did not directly monitor the work of the SMMs, relying instead upon reviewing the Product samples, the QMDRs, and various performance parameters on the BAC1-11 fleet, such as recurring defects and ADDs. The Station Maintenance Manager was directly responsible for the work of the SMMs, but was limited in this role (the AAIB claimed), first, by being the same grade as the SMMs and, secondly, by being rarely present on the night shifts. These assertions were contested by BA management, who indicated that the SMMs were appraised by the AMM, who was well aware of their technical and managerial competence, as well as the

97

quality and quantity of the work produced by their shifts.

Random failure versus systems failure

One of the most interesting features of this accident is that its causes are open to several interpretations. The AAIB posed two extreme alternatives:

- *Random failure theory:* the SMM's errors and violations were a 'one-off' unheralded by any prior lapses or poor working standards.
- *Systems failure theory:* the unsafe acts were typical of the SMM's working practices, which had been observed but condoned or not spotted due to poor monitoring and supervision.

In their published report, the AAIB concluded that the truth lay somewhere between these two extremes: 'The system element being that which accommodated the application of inadequate standards by the SMM for some time, and the perceptual errors contributing the random element.' However, the general tenor of their report indicated that they were mainly in favour of the systems failure theory. To what extent is this view justified?

There is no such thing as a 'gremlin-free' system. All technological operations are beset by latent failures, and the Birmingham maintenance facility was no exception, as the local and organizational factors discussed above testify. The AAIB's main concern, however, was that the SMM's action could only have been symptomatic of a culture that either condoned or was indifferent to poor working practices. And by this they meant a culture in which it was acceptable to exercise judgement rather than following mandated procedures to the letter.

Organizations are like animal species. To survive and prosper, they need to adapt to the local conditions. A major part of the core business of the Birmingham maintenance facility was to keep a fleet of elderly and multiply-modified BAC1-11s both safe and on schedule. To do this successfully, as they mostly did, it was necessary to be flexible and highly committed to the work. Each person in this relatively small and closely knit establishment could see the results of his efforts in aircraft that left the hangar ready for the line. It is not surprising, therefore, that both the morale and the workload were high. The two things go together when we feel valued and can see the direct results of our efforts.

The accident and the investigations it generated provided a

snapshot in time of the Birmingham culture. The picture that emerges is one of competitive pride between shifts, a high degree of job satisfaction and old aircraft generating revenue, yet still for the most part flying safely. Such was the enthusiasm that the local management had repeatedly cautioned the night shifts not to attempt more than was prudent in the time available.

Is this symptomatic of a defective organizational culture? It is if you believe, with 20-20 hindsight, that all aspects of aircraft maintenance can be wholly proceduralized. It is not if you recognize that individuals like the SMM are paid for their professional judgement rather than for their ability to read and obey the procedures. However – and here is the worm in the apple – judgements can sometimes be wrong. But so also can the procedures, as the nuclear power industry has discovered. In a survey of events occurring in 1983 and 1984 (INPO, 1985), the Institute of Nuclear Power Operations identified 387 root causes as responsible for 180 significant events. Of these 52% were classified as Human Factors problems. Of these, 43% were categorized as due to deficient procedures or documentation, especially in maintenance and testing activities. Only 16% of the Human Factors problems were associated with a failure to follow adequate procedures. It could hardly be said that the nuclear power industry has fewer safety critical features than aircraft maintenance.

A question of perspective

Each party to an accident inquiry has a particular perspective: the investigators, the airline management, the regulators and the people directly involved. All are subject to hindsight bias and each group will be inclined to put their own slant upon both the causes and their implications. In this concluding section, we shall add yet another view: a Human Factors perspective. But this, too, will be a personal view and one not entirely in agreement with the opinions expressed by the Human Factors expert who formed part of the AAIB investigating team.

Something that (in our view) counted strongly against the systems failure theory, favoured by the AAIB, was the fact that the SMM's errors and violations formed a coherent series, with each one leading to the next rather than being a collection of isolated incidents. Had the misinstallation of the window been due to a set of discrete lapses and bad practices, involving separate tasks and different people, then the case for the systems failure theory would be much more convincing. As it was, the inter-related and sequential nature of the unsafe acts

99

strongly suggested a 'once-in-a-lifetime' episode. As mentioned earlier, one of the basic principles of error management (as well as being a corollary to Sod's Law) is that the best people can often commit the worst errors. It is also the case that supervisors are represented more often in accident sequences than their numbers would justify on a purely chance basis.

But that is not the whole story. One of the most important lessons of this accident is that organizational characteristics that are highly valued in most circumstances – that is, high morale, a keenness to keep the show on the road and a flexible approach to problem solving – can, in certain rare situations, also be a formula for disaster. And this was one such situation.

Faced with a near-catastrophe, we naturally expect the causes to be rooted in something equally bad, such as the kind of sloppiness that Mr Justice Sheen found in the shore-based management of the *Herald of Free Enterprise* (see Sheen, 1987). But no such indications were present here, though the AAIB tried hard to unmask them. On the contrary, the whole episode began because the SMM did not like to fix windscreens with dirty or scratched bolts. He cut procedural corners because – unwisely as it turned out – he did not feel them to be justified in the circumstances. He knew what had to be done, and believed he had identified the means to do it (the 7Ds). In 999 occasions out of a thousand, his intended actions would have got the job done without any adverse outcome. The exercise of professional judgement is, like it or not, the everyday reality of aircraft maintenance, and will continue to be until the job is done by robots.

It is very natural for all the other parties to the investigation – the AAIB, BA management and the CAA – to condemn this use of professional judgement, particularly when it turned out badly in this instance, and to advocate strict adherence to procedures or the use of additional checklists. But, as other industries have discovered to their cost, more procedures and stricter enforcement are not universal panaceas. To be effective, procedures have to be correct, available, credible and appropriate for the situation at hand. And that is a very difficult act to achieve.

Conclusions

This accident arose from a highly unlikely combination of circumstances. It was one of Sod's more imaginative creations. The important lessons to be learned from it, we believe, are the following.

Such accidents will recur at highly infrequent intervals so long as people, with all their strengths and weaknesses, are employed to maintain aircraft in a commercial context.

They cannot be prevented by simple prescriptions: more procedures, more enforcement, more defences, more checking, and the like. Indeed, such counter-measures could themselves turn out to be implicated in some future accident. (*Note*: Take-off monitors, designed to prevent a recurrence of the Munich runway accident in 1956, were instrumental in causing the Comet crash at Ankara some years later. . . and this is but one of many cases in which the supposed remedy for one accident caused another.)

The best way of minimizing the likelihood of such accidents (we will never prevent them entirely) is to make regular checks upon the situational and organizational factors that promote errors and violations, and to fix those things most in need of repair. There are no 'magic bullet' solutions to forgetfulness, momentary inattention, corner-cutting or perceptual confusions. These psychological factors are the last and the least manageable links in the accident chain. In general, it is much more efficient and cost-effective to tackle the local and organizational circumstances which promote them.

References

Air Accidents Investigation Branch (1992) *Report on the Accident to BAC One-Eleven, G-BJRT over Didcot Oxfordshire on 10 June 1990*. London: HMSO.

Bacon, F. (1620) *The New Organon*. (Edited by F. Anderson). Indianapolis: Bobbs-Merrill, 1960.

INPO (1985) *An Analysis of Root Causes in 1983 and 1984 Significant Event Reports*. INPO 85-027. Atlanta, GA: Institute of Nuclear Power Operations.

Reason, J. (1990) *Human Error*. New York: Cambridge University Press.

Sheen, Mr Justice (1987) *MV Herald of Free Enterprise. Report of Court No. 8074*. London: Department of Transport.

5 The Australian airmiss study

Introduction

This chapter describes a major study of the safety of the Australian Air Traffic Services (ATS) system which utilized the Reason model as a conceptual basis for the investigation. The study was carried out by the Australian Bureau of Air Safety Investigation (BASI).

At the outset, it should be emphasized that the investigation was carried out with the full cooperation of the management and line staff of the Australian Civil Aviation Authority (CAA), the agency which is responsible for the provision and management of air traffic services in Australia. During the project, and since its completion, the CAA has continued to implement changes to rectify systemic deficiencies identified both in this investigation and in a separate comprehensive review of the Australian ATS system to improve further the safety and efficiency of Australian air traffic services.

At the time of writing, these changes included the establishment of a high level Strategic Planning Unit, the upgrading and strengthening of the ATS quality assurance function, the introduction of a Human Factors training programme for all air traffic services personnel, and the use of the Reason model in ATS occurrence investigation and systems safety.

Background: the Australian Bureau of Air Safety Investigation (BASI)

The Bureau of Air Safety Investigation is the agency of the Australian

102

Government responsible for the investigation of accidents and incidents occurring to civil aircraft in Australia and its territories. BASI operates in accordance with Annex 13 of the Chicago Convention, (ICAO) and Australia has incorporated the provisions of this Annex into its domestic Air Navigation Regulations.

The objective of the Bureau is to promote safe aviation by disseminating information and safety recommendations based on the investigation of selected accidents and incidents, and on the results of research.

The results of BASI investigations are presented in terms of 'findings' and 'significant factors'. The Bureau does not utilize the term 'cause', nor does it, as some other countries do, identify any one factor as the most important in a particular occurrence. In the Bureau's experience, to nominate any one factor – for example, as a 'most probable cause' – results in most attention being concentrated on that factor alone, when the majority of air safety occurrences are the result of a complex interaction of many factors. Every one of these should be addressed for purposes of prevention, otherwise the effectiveness of the investigation process can be greatly reduced.

In practice, the Bureau does not distinguish operationally between accidents and incidents – they are all air safety occurrences. BASI has developed and refined various criteria to decide which events will be examined most closely – one of these is a primary emphasis on the safety of fare-paying passengers in any category of operation – Regular Public Transport (RPT), charter and commuter.

The Bureau is placing greater emphasis on applied research and analysis. The objective is to identify underlying factors within the aviation system which can impact upon safety at the 'sharp end' – that is, in the cockpit, cabin, control tower, maintenance workshop, or on the ramp.

Contemporary aviation safety is no longer primarily the province of pilots and engineers alone – it is now a multidisciplinary endeavour, often requiring input from many diverse fields. The aviation industry is going through a process of rapid change in its operational, technical, commercial, economic, regulatory and political dimensions. Developments in each of these dimensions can impact directly and/or indirectly on air safety.

Reflecting the increasingly multidisciplinary approach of its operations, Australia implemented, in late 1992, a single employment category of 'air safety investigator', within which there are at present six different types – operations (pilots), air traffic services (ATS), human performance, engineer, maintenance (LAME) and technician (e.g. electronics, communications).

BASI employs a team approach to investigations whenever possible. For example, the investigation of a serious ATS incident could result in the assembly of a team of investigators in the categories of air traffic services, operations (pilot) and human performance. All work together as an integrated group to apply their different fields of professional expertise to the identification of the significant factors in the occurrence.

Perhaps the single most important change in the Bureau's recent history is the shift from a primarily reactive organization, investigating accidents and incidents after they occur, to an organization which is also proactive and equally concerned with the prevention of air safety occurrences. BASI now dedicates considerable resources to the proactive identification of deficiencies in the aviation system which have the potential, given the right combination of events and circumstances, to become significant factors in accidents.

Similarly, for the Bureau's 'reactive' investigations, as well as finding out what went wrong at the 'sharp end' – in the cockpit, control tower, cabin or maintenance area – BASI aims to identify any underlying factors in the aviation system which might have contributed to the occurrence.

For example, if passengers are killed or injured because of problems of evacuation following a survivable accident, it is of course essential to determine what actually went wrong in the cabin, and to ascertain the immediate reasons why the crew were unable to ensure the survivability of all occupants. But of even greater importance is the identification of any latent organizational factors which may have contributed to the failure of the cabin crew to achieve a completely successful evacuation – factors such as poor communication, inadequate supervision or inappropriate training.

Experience has shown that if such underlying systemic factors are not detected in investigations, and therefore remain unchanged, similar system breakdowns will inevitably occur again. The Bureau's new Safety Analysis Branch, which has been based on the successful Canadian Transportation Safety Board model, is dedicated to the critical task of identifying these underlying latent deficiencies in the aviation system, both before and after accidents or incidents.

For many years, a characteristic of the aviation industry has been that, although the world airline accident *rate* is very low, it has effectively plateaued. If this rate does not change, and the projected increases in air traffic occur as predicted, then early in the next century the *number* of airline accidents will increase to a level which will be completely unacceptable to the travelling public – approximately one hull-loss every fortnight (Freeman and Simmon,

1990).

It is therefore imperative that if a breakaway from this steady state is to be achieved, and substantial improvements in what is already an excellent safety record are to be realized, a fundamentally new approach to air safety is required. The industry cannot afford simply to recycle its present well established safety philosophies and procedures. It must do more to achieve significant change to the *status quo.*

Fundamental to the Bureau's new philosophy is the conceptual and theoretical approach to the safety of complex sociotechnical systems developed by Reason, described earlier in this book.

The Bureau has adopted the Reason model as the basis for its investigations, and this chapter describes a major systemic ATS investigation using this approach.

The airmiss study: introduction

The Australian Air Traffic Service (ATS) system is provided and managed by the Civil Aviation Authority (CAA). It is intended to, and does, provide a safe, efficient and cost-effective air traffic control and advisory service to the aviation industry within domestic and international (oceanic) airspace.

As part of BASI's proactive approach to aviation safety, a programme of active monitoring of certain systemic safety health indicators was initiated. One such indicator was that of ATS-related occurrences. In this regard, BASI's aim was to gain deeper insights into the functioning and safety 'fitness' of the ATS system through the utilization of a systemic approach to the investigation of ATS occurrences.

Methodology

Many ATS incidents have the potential to reveal latent failures within the ATS system. The vulnerability of the ATS system to both human and technical failure is perhaps most clearly revealed through airmiss occurrences.

Airmiss reports are classified by BASI according to the risk of collision between the aircraft involved. The most serious incidents are those which involve a high risk of collision with the aircraft passing within 100 ft vertically and 500 ft horizontally.

The airmisses, which were taken as the primary basis for the analysis, occurred in June and July of 1991. At that time the CAA was undergoing considerable change, and there was significant uncertainty regarding the future structure and technology of air traffic services in Australia.

In June and July of 1991, 31 reported incidents were classified as airmisses – that is, occurrences in which there was the potential for collision between aircraft. In deciding the level of investigation to be dedicated to any occurrence, the Bureau also takes into account the potential safety benefits which may be gained from the investigation. Consequently, the risk of collision may not be serious in each case, but the types of systemic deficiencies, of which the incident might be a token, could be significant.

This BASI investigation was conducted during 1992. Its aim was to identify and understand the organizational and system characteristics, together with the unsafe acts, which lead to certain kinds of ATS occurrences. One of the Bureau's primary objectives was to provide the CAA with safety recommendations which would enable the significant systemic factors to be remedied at their origins – in other words, 'to drain the swamps' (Reason, 1993). A secondary, but very important, aim of the investigation was to demonstrate the practical safety benefits of the application of the Reason model to the ATS system.

Eight of the 31 airmisses were identified by the Bureau as being serious occurrences. The reasons for the selection of those incidents were that they:

(a) involved regular public transport (RPT) aircraft which carried fare paying passengers;
(b) occurred in controlled airspace; and
(c) involved actual/potential breakdown in separation standards.

Detailed summaries of the eight primary incidents involved are provided in the Bureau's published report. These eight incidents, although the primary basis of the analysis, were supplemented by other airmiss occurrences to aid the analytical process. A complete listing is provided in the BASI report (BASI, 1994a).

Application of the Reason model to air traffic services

In the following description, the Reason model has been adapted to the ATS system.

Unsafe acts

The ATS system is ultimately reliant upon the human operator to

106

process air traffic in a safe and expeditious manner. For the most part, this is carried out safely. However, human performance is fallible, and on occasions controllers and pilots commit unsafe acts.

When considering such acts, a distinction can be made between violations and errors. The categorization is based on whether the unsafe act was intentional or unintentional. Such categorization is useful when considering the unsafe acts perpetrated by both controllers and pilots.

Errors Reason (see Chapter 1) describes two distinct types of error:

(a) *attentional slips* and *memory lapses*, which involve unintended deviation of actions from what may be a perfectly good plan; and
(b) *mistakes*, where the actions follow the plan but the plan deviates from some adequate path to the desired goal.

Air traffic control is based upon the processing of information provided in aural, visual or written form. In this dynamic and complex cognitive and operational environment, it is understandable that occasionally controllers do not, or are unable, to consider fully the ramifications of some element of the traffic pattern, or they may misapply some rule, despite their high degree of training and the standardization of the procedures they use.

The majority of unsafe acts which occur within the ATS system are likely to be categorized as errors.

Violations Violations are deliberate deviations from regulated codes or procedures. Such acts sometimes occur within the ATS environment. Reason (see Chapter 1) describes three types of violations:

(a) *routine violations*, involving short cuts between points within a task;
(b) *optimizing violations*, in which the individual seeks to optimize some goal other than safety; and
(c) *exceptional violations*, one-off breaches of regulations seemingly dictated by unusual circumstances.

Probably the most common kinds of violation occurring within any ATS system are those whereby controllers might attempt to make the system more efficient, or temporarily increase the traffic capacity. For example, a controller might deliberately allow the formal published aircraft separation standards to be infringed for a short time if he/she does not believe safety will be compromised.

Summary The error/violation categorization is useful in assessing unsafe acts made by controllers, as remedial action will differ according to the type of error or violation. The imperfect cognitive functioning associated with errors should not be categorized in the same way as violations. Violations have a motivational basis and can only be properly understood in an organizational context.

The incidence of violations can be most effectively reduced by changing those organizational factors which most directly affect particular local factors such as attitudes, group norms and morale. The probability of errors may be reduced by the organization addressing specific local factors such as training, workplace and task design, time pressures, and environmental elements such as lighting and noise.

Psychological precursors of unsafe acts

Within an organization, psychological preconditions may exist which increase the likelihood of unsafe acts. Some may be directly related to the nature of the organization itself; others may be a product of the worker's private life. Examples of such preconditions which could exist within the ATS environment are: insufficient or excessive workload; poor human-system interface; conflict between management and staff; group norms which condone violations; a culture which encourages risk-taking; and, disturbed sleep patterns.

Organizational deficiencies (line management decisions)

Reason states that decisions made by management over a long period of time may have created certain inherent flaws within an organization. Such decisions may be influenced by a lack of information or resources, time pressures, higher level decisions, enforced decisions brought about by restructuring, and so on. The consequences of these decisions may take considerable time to manifest themselves, and in most cases are evident in the psychological preconditions referred to above.

The interaction between organizational deficiencies and psychological precursors of unsafe acts may be complex. An example used by Reason (see Chapter 1) illustrates this point:

> Deficiencies in the training department can manifest themselves as a variety of preconditions: high workload, undue time pressure, inappropriate hazard recognition, ignorance of the system and motivational difficulties.

108

An earlier and comprehensive review of the Australian ATS system carried out by Ratner and associates (Ratner, 1992) identified organizational deficiencies which could affect safety in the transition to completion of the Australian Advanced Air Traffic System (TAAATS). This is a major project intended to substantially upgrade the Australian ATS system, utilizing highly advanced technology. Examples of organizational deficiencies identified by Ratner were: an inadequate ATS Quality Assurance function; lack of documentation and staff training in the operation of a revised 'two tier'[1] safety regulation and surveillance scheme; and, little formal accountability for safety at a managerial level

Corporate actions (senior management decisions)

Actions of the most senior management and the board of the CAA, like those of any large organization, have the potential to impact, however indirectly, on the actions of the ATS operators at the workface. Decisions regarding future directions of the organization, allocation of resources, and even publicly stated goals, all impact upon workers.The ramifications of such decisions may lie dormant within the system for years before some combination of events and circumstances exposes fallibilities within those decisions.

Inadequate defences

A properly designed system has in-built defences to protect it from human or technical failure. The ATS system has a number of these defences. Examples include: readback of instructions; position reports; single direction routes; standard aircraft cruising altitudes; Standard Instrument Departures (SIDs), and many others. Such provisions ensure that the potential for error on the part of either a pilot or a controller is minimized. At present, the majority of ATS system defences are based upon the ongoing comprehensiveness and veracity of the individual controller's awareness of the present and future traffic situation, in other words, the accuracy of his/her mental model of the 'big picture'.

---

[1] The 'two tier' policy was a result of the CAA restructuring its safety regulation and surveillance activites from the traditional system of direct monitoring in the field '...to one that requires the aviation industry to undertake some of the safety related activities in the areas of airworthiness, training standards, and operational supervision programs that were previously done by CAA. CAA's role becomes that of monitoring industry programs and their management, rather than monitoring airmen and aircraft directly.' (Ratner, 1992, p 24)

ATS occurrences – a tragic example

In practice, only a very small percentage of unsafe acts lead to an occurrence which is ultimately detrimental to the safety of the system. In most instances the multiple layers and redundancies of the ATS safety defences act effectively to protect the system – for example, if a controller clears the wrong aircraft to a higher altitude, the pilot may recognize that the clearance is not appropriate. In addition, the controller may realize his/her error when reviewing the flight strips, or when the pilot reads back the clearance.

In instances where the layers of defence are breached, a number of factors must occur in conjunction to produce an incident, or in some extreme cases, a catastrophic accident.

An example of such an accident occurred at Los Angeles International Airport in February 1991, when a Boeing 737 collided with a Metroliner. The Boeing had been cleared to land on a runway on which the Metroliner was lined up awaiting a take-off clearance.

The accident occurred at night, and the lights of the Metroliner were indistinguishable from all the other lights associated with the runway and taxiways.

A number of factors contributed to this accident: there was confusion on the part of ATS personnel over the call signs of several Metroliner aircraft which were simultaneously manoeuvring on the airport: the view of the runway threshold from the control tower was obstructed; and the flight strip for the Metroliner involved in the accident was missing.

In this accident, the local controller did not maintain an adequate awareness of the overall traffic situation, which culminated in the inappropriate landing clearance. This occurred at a time when the Metroliner's conspicuity to both the Boeing 737 aircrew and the tower cabin personnel was significantly reduced. When this accident scenario is reviewed, it is apparent that the controller's error could have been detected on numerous occasions by a number of different people – for example, the controller, the Boeing 737 pilots, the pilot of the Metroliner involved, or the crew of another Metroliner which the controller had confused with the aircraft involved in the accident.

The National Transportation Safety Board (NTSB) determined that the probable cause of the accident was:

> ...the failure of the Los Angeles Air Traffic Facility Management to implement procedures that provided redundancy comparable to the requirements contained in the National Operational Position Standards and the failure of the FAA Air Traffic Service to provide adequate policy

direction and oversight to its air traffic control facility managers....
Contributing to the cause of the accident was the failure of the FAA to
provide effective quality assurance of the ATC System. (NTSB, 1991, p
76)

Incident data as a source of information for systemic analysis

Both individual occurrence and aggregated data from ATS incidents
have the potential to provide information regarding the functioning of
the many components of the ATS system. Perhaps the most
immediately apparent benefits are the appreciation of the nature of
the active failures of controllers and pilots, and how the safety
net/defences of the ATS system can be breached. However,
investigations of such occurrences can also provide insights into latent
organizational failure types within the ATS system.

Using incident data to assess the contributions made by the various
levels in the Reason model may be difficult. The chain of causality in
complex organizational systems may itself be complex and subject to
ambiguity. In a single incident, what one observer might perceive as
clear evidence of conflicting organizational goals, another may view as
after-the-fact rationalization of error. Considering a number of
incidents as a group helps to resolve some of these difficulties. By
combining data from a number of investigations, underlying patterns
are more likely to become apparent (see the ICAO CFIT safety analysis
discussed in Chapter 6). Consequently, a better and more robust
understanding of the underlying organizational factors may be gained.

Historically, a theoretically structured systemic approach has not
normally been applied to accident investigations at the outset. Readers
of accident reports expected to be given a clear logical connection
between the factors contributing to the accident and the accident
itself. With regard to those local factors immediately proximate to the
accident, such an approach remains appropriate. This element of the
analysis is well understood and documented in the International Civil
Aviation Organization (ICAO) Accident Investigation Manual (ICAO,
1970).

However, the broader systemic approach to air safety investigation,
which, in addition to determining local factors, also aims to identify
and remedy in a principled way the organizational factors which
facilitate safety occurrences, has only in comparatively recent times
become accepted. O'Hare (1994) describes this development as a
'...quiet but steady revolution...' (p4).

Analysis of ATS safety: the BASI airmiss study

This section describes the Bureau's investigation and analysis of the safety of the ATS system using the Reason model.

Safety deficiencies are identified and are illustrated with specific incidents for each component of the model. The components under consideration are: inadequate defences; unsafe acts; psychological precursors (error-producing conditions); and organizational deficiencies (latent organizational failures). Actions which could be taken to rectify the deficiencies identified are also described. A later section outlines some of the remedial actions which the CAA has taken in response to the investigation.

Inadequate/failed defences

Human or technical failures in the ATS system are not infrequent. However, when an error or equipment breakdown does occur, the system should be immune to the impact of the event by detecting and rectifying the anomaly before it contributes to an occurrence. The present analysis revealed that the defences of the ATS system were inadequate. There were few procedures or technological aids which protected the integrity of the system against controller error and technical failure (see Figure 5.1).

The safety of the ATS system is dependent to a considerable extent on error free performance by controllers. When an error is made, such as an incorrect clearance, in the majority of cases only the sector controller will be fully conversant with the situation, and therefore will be the only ATS operator in the system who is able detect the error. At the time of this investigation, aircrew seemed to be unwilling to scrutinize fully controllers' actions. Thus, the vital role which pilots can play in detecting error is under-utilized.

Examples of technical defences available to ATS officers overseas at

---

- Controller must act as own safety net
- Aircrew role not emphasized/understood
- Inadequate monitoring
- Inadequate verification or validation of data
- System has few 'failsafes'
- Collision avoidance not planned

Figure 5.1.   Inadequate defences identified

the time of the investigation were 'conflict alert', or 'conflict probe'. An airborne system such as Traffic Collision Avoidance System (TCAS), which has the capability to detect potential conflicts, was becoming increasingly available to pilots. At the time of the study, the Australian ATS system had few failsafe sub-systems, and even in places where failsafe systems could and should have existed, the Bureau's investigation identified elements which were 'fail unsafe'.

For example, the flight data processing errors discussed later indicated the problems associated with manually transcribing information onto flight strips, and then transferring that information to controllers who had no way of validating the information so provided. In other words, there were no specific defences to protect the ATS system from the consequences of errors in flight strips.

Corporate direction

A fundamental premise of the Reason model is that the corporate level has a vital role to play in determining the safety health of an organization. The present analysis of the safety health of the CAA ATS division identified some key deficiencies. While BASI recognized that the CAA had gone a considerable way in its attempts to rectify problem areas, the Bureau believed that the Authority had further areas to address. These are shown in Figure 5.2.

The ambiguity between safety and service which existed at the workface can only be removed by the promulgation and promotion by management of a clear organizational safety philosophy which clearly defines the balance between service and safety. In addition, the CAA needed to address the requirement for improved defences for the ATS system.

- Development of a clear safety philosophy
- Clarification of service versus safety trade-off
- Preparation of a national strategic plan for Air Traffic Management
- Development and implementation of an integrated approach to training
- Establishment of an effective Quality Assurance (QA) function
- Provision of adequate defences to increase the error tolerance of the ATS system

---

Figure 5.2. Areas identified as requiring action

113

Initiatives in the spheres of Air Traffic Management and training had been introduced. However, there remained a need for continued management support to ensure the planning, implementation and consolidation of such programmes. Such high level support was also necessary to guarantee the effectiveness of the Quality Assurance (QA) function.

Unsafe acts

During the course of the investigations, it was apparent that the errors or violations which result in air safety incidents are numerous and wide ranging. This situation is not unique to Australia. Figure 5.3 summarizes the most significant unsafe acts identified in the study.

Planning Air traffic control involves a continuous process of devising 'game plans' to meet current and future traffic requirements, along with assessing and reassessing those plans and making adjustments as events unfold. For example, the plans should allow for an aircraft failing to clear a particular flight level in time, and should ensure the required traffic separation. An unsafe act can occur if the controller does not have a contingency plan for dealing with eventualities. It is worth noting that in Canada, the ATS authority, Transport Canada, defines a planning failure as a reportable operational incident.

In the present study, when planning failures were considered in depth there seemed to be a focus on solving immediate problems, and to somehow 'get by' despite the situation. It was found that this lack of a defensive posture may even occur in low workload situations, when at the initiation of the traffic sequence the various possible contingencies are not assessed.

- No, or inadequate, plan for traffic processing
- Excessive reliance on 'expected aircraft performance' or 'aircraft performing as anticipated'
- Failure to maintain the traffic picture (situational awareness)
- Inappropriate use of flexibility to vary procedures
- Providing service without checking outcomes
- Inattention to primary task
- Coordination failures
- Flight data processing errors

Figure 5.3. Unsafe acts identified during incident investigations

114

A significant category of unsafe act was the controllers' reliance on expected aircraft performance during the formulation of their traffic management 'game plans'. It is understood that all controlling is based on expected outcomes of instructions to aircrews and aircraft. However, an integral characteristic of good control is being continually alert to the possibility that such expectations might not always eventuate. Safety within the ATS system is maintained by planning to ensure that an 'escape hatch' is available should an aircraft fail to perform as anticipated, and by providing for timely adjustment of the plan to take account of the actual performance of the aircraft.

There were indications that on some occasions controllers relied too heavily upon expected aircraft performance, and failed to monitor the traffic situation adequately as it unfolded (see Inserts 1 and 2).

Situational awareness   Continuous awareness of the traffic disposition is an essential element of air traffic control. For various reasons, such as distraction, this situational awareness may sometimes be degraded. In most instances, such degradation will have little overt impact on traffic processing as controllers are able to quickly rebuild their 'mental model'of the traffic pattern by reference to radar information, or by reviewing the flight progress strips. However, it is possible for degradation of situational awareness to reach a level at which it can contribute to a breach of separation standards.

Controllers have on occasions overlooked or disregarded the presence of another aircraft under their jurisdiction (see Insert 3), or one which had been recently transferred to another sector (see Insert 4). A number of specific investigations which were reviewed by BASI identified an apparent association between lack of situational awareness and incomplete monitoring. When the controller's attention was directed back to the unfolding situation, minimal time was then available to effect a satisfactory resolution of the problem.

Flexibility to vary procedures   Some degree of flexibility is built into the ATS system to allow traffic to be processed in the most efficient manner possible. However, if this flexibility is used to excess, or at an inappropriate point in time, then safety standards may suffer. Insert 5 provides an example of a situation in which the original departure clearance was modified on two occasions, resulting in differing expectations on the part of the Departure cell and the Tower.

Coordination   The effective transfer between sectors of information regarding aircraft is vital to the safe and efficient operation of the ATS system. For this reason, coordination between sectors is bound by

---

Insert 1   B/911/3134   **BRISBANE**   6th June 1991

This occurrence was reported as a breakdown of radar separation standards within the terminal area between a Cessna C210 (C210) conducting an Instrument Flight Rules (IFR) training flight and a McDonnell Douglas DC10 operating an international Regular Public Transport (RPT) flight. Recorded radar data indicated that the aircraft passed with less than one nautical mile horizontal separation when they were at approximately the same altitude of 3700 feet.

The DC10 had departed Brisbane on a Standard Instrument Departure (SID). The C210 departed Archerfield for Maroochydore and had been cleared to climb to 6000 feet without restriction on the direct track, by the approach (APP) Controller. The DC10 failed to commence its turn at the point specified in the SID. This was brought to the attention of the Approach controller by the tower controller.

APP then turned the C210 onto a heading of 270 degrees and DC10 onto a heading of 340 degrees. The instruction required the C210 to turn towards high terrain, although the controller did not know whether the aircraft had sufficient height to clear the terrain. Once the aircraft flight paths had diverged, the C210 was then turned back onto its original track. This may have resulted in the C210 flying through the wake turbulence of the DC10.

The investigation found that the controller, assuming that the DC10 would follow the SID, failed to adequately monitor the actual flight path. The APP controller had no other traffic for processing at the time of the occurrence. There were no traffic capacity problems restricting alternative vectoring options. However, other options may have involved additional coordination.

---

procedures regarding when and how such information should be transferred. In a system so dependent on the accurate transfer of verbal information, it is not surprising that controllers occasionally fail to recognize the implications of coordinated information, such as advice that both aircraft were operating on the same track, or that departure instructions had been amended in part or in full (see Insert 5). In such instances, the expectation of traffic movement can be different for each contiguous controller. If the initial error is not detected in subsequent coordinations, a breach of the system's defences can occur because of technological limitations. Insert 6 is an example where a controller did not detect that two aircraft were

Insert 2   B/914/3071   ADELAIDE   17th July 1991

An Airbus A320 (EA32) departed Adelaide en route to Brisbane. The aircraft was given an unrestricted climb to flight level (FL) 370 via air route T77 to Brisbane. Adelaide control were asked by Melbourne to ensure that the aircraft was at FL330 by 20 nautical miles (nm) south west of Mildura as a Boeing 747 was tracking on a crossing route at FL310.

A Boeing 767 was flying from Sydney to Adelaide via Mildura maintaining FL310 on the reciprocal heading to the EA32, and appeared on the Adelaide control radar approximately 124 nm from Adelaide.

Adelaide Sector 4 (SEC 4) contacted the EA32 and requested that the aircraft maintain best rate of climb to FL330. The objective was to climb the EA32 above the B767, using a radar standard.

Approaching FL310, the EA32 experienced an increase in ground speed, increasing the closing speed with the B767. Following this the airspeed of the EA32 fell below the minimum manoeuvring speed and the Captain reduced the climb angle to accelerate the aircraft. By the time the aircraft had passed, the EA32 had regained the best rate of climb; however the applicable separation standard had been breached.

The SEC 4 controller had monitored the climb of the EA32 until the aircraft was approximately 40 nm west of the B767. At the displayed climb rate, he believed that the EA32 should have been at FL330 by the estimated time of passing. He had then turned his attention to another aircraft. When he returned to the EA32/B767 he noticed that the radar returns had merged. The aircraft had passed at approximately 110 nm east of Adelaide with less than the required separation standard.

operating on the same route.

Service without checking outcomes Air traffic controllers provide a 'service' to the aviation industry. However, in providing this service safety standards may be undermined if controllers fail to check the outcome of their instructions. This type of unsafe act is characterized by the rapidity of controllers' responses to requests from aircraft, such as for permission to deviate from a previously submitted flight plan. If a controller responds immediately, it is probable that a complete assessment of the effect and outcome of the resultant change to the traffic situation arising from acceding to the request could not have

117

Insert 3 B/911/3141   CAIGUNA   12th June 1991

In this occurrence two jet RPT aircraft were operating on the same one-way route in procedurally controlled airspace when the controller approved a higher flight level request from the following, faster aircraft. This resulted in a breakdown of procedural separation standards.

Both aircraft in this occurrence had initially been cleared to climb to flight level 370. The Sector controller requested that the Arrivals controller modify the flight level of the second aircraft, a Boeing 747-400. The Boeing was therefore recleared to FL350.

While both aircraft were under the jurisdiction of the Sector controller, he assisted another Sector controller with plotting separation standards for two other aircraft. When the crew of the B747-400 requested the availability of FL390, the controller said 'affirm descend to FL290 ... correction was that FL290 or 390?'. The crew responded 'three nine'. The controller immediately cleared the aircraft to FL390, failing to recognize the significance of the level change request, nor the proximity of the other aircraft at FL370.

Insert 4 B/913/3158   EILDON WEIR   18th July 1991

An Airbus A320 (EA32) and a Boeing 727 were operating scheduled domestic RPT flights to Melbourne with arrival sequencing being conducted in the vicinity of Eildon Weir (ELW). During the subsequent vectoring, a breakdown of separation standards occurred between the EA32 and the Boeing 727.

In this instance, the Boeing had been transferred to the Arrivals (ARR) controller and had reduced speed as requested to 230 kts. The EA32 was required to enter the holding pattern, and it was instructed to turn onto a converging course, towards the B727, then descending to FL160. The radar screen labels available to the controller included a readout of aircraft level and ground speed. These would have provided evidence that the closing speed between the two aircraft was approximately 180 kts. However, the proximity of the two aircraft was only brought to the attention of the Sector controller when the EA32 was instructed to contact ARR. The crew acknowledged the instruction and requested the level of the preceding aircraft. The Sector controller then realized that separation had been lost between the EA32 and the slower B727.

Insert 5   B/916/3018   SYDNEY   13th July 1991

After the initial departure instructions were given to the Sydney control tower for an Airbus A300 (EA30), the departure sequence was changed twice, with two aircraft sequenced ahead of the EA30. Immediately following the departure of the second of these aircraft (a British Aerospace 146), Departure Radar (DEP) indicated that the EA30 could be unrestricted. The tower understood the instruction to mean cancel the previous departure instruction; however the DEP controller only intended cancellation of the altitude restriction.

Insert 6   B/911/3197   NE PERTH   11th July 1991

This breakdown in separation standards involved two Boeing 737 RPT aircraft which were thought to have been operating on two different (but converging) air routes within controlled airspace. Controllers with jurisdiction for the aircraft were providing separation based on dissimilar and incorrect flight progress strip information. The error remained undetected until both aircraft passed on the same track on the peripheral range of the Perth radar display.

The flight progress strips for the inbound aircraft prepared for Perth Arrivals (ARR) showed the aircraft would arrive via route W43. However, the strips for Perth area control Sector 2 (SEC 2) displayed the correct route T31.

SEC 2 coordinated the arrival of the inbound aircraft with the Perth Arrival Procedural (ARR (P)) controller, based on an estimated time at 160 nautical miles (nm) from Perth. This was the limit of radar coverage. Coordination for an aircraft operating on W43 was also required by SEC2 to provide for an estimate at -160 nm. Neither the ARR(P) controller, nor his trainee, noticed that the route was different to that displayed on the ARR(P) strip.

---

been achieved by that controller (see Insert 3).

Similarly, the apparently high incidence of unrestricted operations and track shortening events (see Insert 1) may indicate that the provision of 'customer service' is paramount, and consequently pilot requests have been accommodated without a complete assessment of their ramifications and appropriate defensive planning.

Inattention to primary task Air traffic control is dependent on each operator being able to undertake a number of different tasks simultaneously in order to develop an integrated traffic processing plan. A further integral aspect of good control is the controller's ability to divide and prioritize his/her attention in an appropriate manner. The fluid and dynamic environment in which controllers work may sometimes lead to situations in which attention is not focused on the most critical elements of the tasks at hand. Controllers may be distracted by the absence of flight progress strips, or by plotting separation standards for other controllers (see Insert 3). In other circumstances, attention to the primary task may also be degraded by noise levels within the working environment, or by discussions on matters unrelated to the traffic situation.

Flight data processing errors Flight progress strips provide controllers with a representation of the expected traffic and its disposition, which enables them to anticipate and identify potential conflicts. Errors in flight strip information may provide controllers with an incorrect mental picture of traffic disposition or expected outcomes. While such errors do not have a primary role in the development of specific unsafe acts by controllers, they may contribute to the development of an unsafe situation. Insert 6 is an example where controllers with jurisdiction for two aircraft on differing but converging routes, were providing separation based on dissimilar and incorrect flight progress strip information. This example, and other cases involving deficiencies in flight data preparation, illustrate how such factors increase the opportunity for unsafe acts to develop and penetrate the ATS system's defences.

Review of findings regarding unsafe acts

Unsafe acts provide evidence of the fallibility of human performance. It can be argued that the controllers involved in the incidents considered in this investigation were products of the system which selected and trained them, and subsequently monitored and maintained their in-service performance standards.

That system had institutionalized attitudes and a culture in its personnel which made it acceptable, and perhaps relatively common, to rely excessively upon anticipated aircraft performance, to overlook separation assurance, to use flexibility to excess, and to work around system deficiencies. The inference is that the ATS system was overly reliant on the error free execution of controller skill.

Although unsafe acts are not uncommon in the Australian, or any other comparable ATS system, they rarely lead to a serious breach of the system's defences because those defences have been developed and proven over time through hard earned experience. As a result they are extremely effective.

It was considered that little would be gained from tackling the local factors described above in isolation. Only by also identifying, understanding and analysing the underlying organizational factors in the ATS system would significant potential for change be realized.

Predisposing psychological factors or error-producing conditions

Predisposing psychological factors are latent states, which affect the potential for the unsafe acts discussed above. Such factors may be viewed collectively as the 'organizational climate' in which errors and violations occur.

The psychological precursors of unsafe acts, sometimes referred to as 'thought influencers' (Charlton, 1993), which were identified in the course of this investigation, are shown in Figure 5.4. They are described in detail as follows.

Excessive self-reliance Controllers are trained to be reliant on their ability to make effective decisions in complex and dynamic situations. At the time of this investigation, this training did not cover the human information processing capabilities and limitations which affect decision making performance. In some instances, excessive confidence in their decision making capabilities has led to situations in which controllers have failed to utilize fully the facilities available to them to alleviate their workload, or reduce the complexity of the traffic processing requirements. For example, controllers seemed

---

- Excessive self-reliance
- Focus on tactical rather than strategic control
- Anticipation used to excess
- Workload (excessive/minimal)
- Acceptance of frequent distractions in the work environment
- Ambiguity regarding the service/safety trade-off
- Work around system deficiencies
- Uncertainty regarding future (1991)

Figure 5.4. Predisposing psychological factors identified during the investigations

unwilling to route an aircraft through another controller's sector. This was particularly prevalent in terminal areas (see Insert 1). Similarly, controllers appeared reluctant to request assistance in cases where the processing demands of the traffic configuration increased in difficulty and overwhelmed the capacity of the sector and the controller.

Focus on tactical rather than strategic or defensive control The second psychological precursor identified was the focus on tactical rather than strategic, or defensive, control. Planning is the basis of all air traffic control, with controllers being trained to anticipate and plan for potential traffic conflicts. However, there seemed to be a tendency for controllers to act in a reactive mode, and to solve problems as they occurred (see Insert 1), with little planning effort directed towards solving 'what if' questions at the initiation of a traffic sequence.

Anticipation used to excess Controllers are frequently required to base plans on their expectations of aircrews' actions, aircraft performance, and the actions of other ATS personnel. In the majority of cases, these judgements or 'gambles' are correct. However, the negative dimension of this situation is that it may increase the likelihood that the controller will rely too heavily on his/her expectations always being realized when making decisions.

The role of aircraft performance in the incidents under review has been discussed previously, as have the coordination problems in which controllers incorrectly assume that they each have the same mental picture of the traffic disposition.

Workload Ratner (1987), in his first review of the Australian ATS system, noted that:

> Human errors are more likely to occur during certain kinds of operational situations, such as those of high traffic complexity and level, and very low traffic levels, and circumstances where coordination is complex. (p 25)

All these situations were reflected in the occurrences considered in the present investigation.

A simple direct correlation between air traffic density and controller workload does not exist. Factors such as experience, foresight, procedures and working environment also play a critical role. High traffic density combined with a number of peripheral tasks may result in a controller being unable to assess the traffic pattern adequately. As a result, the controller may fail to recognize a potential conflict. In other situations, the level of traffic may not be extreme, but the configuration of the airspace or the physical nature of the controller's workstation may result in an overall level of task complexity which reduces the controller's ability to perform with minimal error. Limited traffic levels may lead to situations in which the controller's full attention is diverted from the primary task of controlling traffic to other ancillary activities – for example, plotting separation standards for another controller, or discussing industrial issues (see Inserts 1 and 3). In low stimulus environments, the maintenance of attention and vigilance is difficult.

Acceptance of frequent distractions in the workplace As indicated previously, diversion of a controller's attention may have occurred prior to an airmiss occurrence. Distraction within the workplace may be responsible for such inappropriate division of attention. If the occurrences under review were representative, it is apparent that undertaking supplementary tasks for fellow controllers, or conversing with other controllers during their time at the console, was routine. Equally, noise levels could make it difficult to concentrate, particularly during a shift change (a factor previously identified by Ratner, 1987).

Indications were that 'on the job training' (OJT) unintentionally introduced distraction and additional workload, both of which were associated with the checking of the trainee's actions and the requirement for explanation. In some instances this led to situations in which the trainer was unable to see the 'whole picture'. OJT also imposed an element of distraction for other ATS personnel working in association with the trainee/trainer combination.

Instances were identified where:

(a) ATS personnel found it necessary to clarify coordination details. The trainee and trainers failed to recognize the ramifications of the information provided – for instance, that both aircraft were on the same air route.
(b) ATS personnel took on additional tasks such as plotting separation for a training combination, which diverted attention from their primary tasks.

While the Bureau did not wish to imply that controllers should work in a sterile environment, in which assistance to others and discussion does not take place, it is believed that the frequency and magnitude of distractions is detrimental to the essentially cognitively based task of controlling aircraft. It is therefore essential that management acts to limit the amount of distraction in the workplace, as documented in standard operating procedures.

Working around system deficiencies Aircraft control techniques and procedures are necessarily adapted for the particular ATS environment. It was found that controllers were forced to adapt their modes of operation to cope with deficiencies which were inherent in these environments. These ranged from inadequate ergonomic design of work stations, poor quality of radar systems and displays, the limits of VHF range, to the non standard phraseology used by some controllers. Other deficiencies included the limited capacity of air routes because of inadequate radar facilities, and the increased complexity of traffic processing because of the route structure and the design of sectors.

Some of these deficiencies are thought to impinge directly upon the potential for unsafe acts by increasing such dimensions as workload and coordination complexity. Others, such as the quality of the radar or the physical working environment, may impact more subtly upon staff morale and motivation, which may also affect the prevalence of unsafe acts.

In this instance the psychological precursor of unsafe acts is the general acceptance by controllers of the requirement to work around inappropriate design features of the ATS system. As traffic increases in numbers and complexity the 'work arounds' themselves may become serious safety deficiencies.

Ambiguity regarding the service/safety trade-off The operation of all complex systems involves trade-offs between service and safety. In the Australian ATS system this involved a trade-off between providing an economically viable service while maintaining safety.

In some circumstances, the balance which the CAA wished to achieve between service and safety was not clearly communicated to all personnel. At the time of this study, BASI considered that such ambiguity existed, and was reflected in instances where controllers, not wishing to inconvenience aircrews, or with the aim of shortening track miles, inadvertently reduced aircraft separation standards below the specified minima.

The service ethos was also demonstrated in the speed at which controllers provided clearances following requests for altitude changes. As previously indicated, the rapidity of response was such that complete assessment of the anticipated outcome could not have been achieved.

During the study period 1991–1992, the organizational balance between the provision of service and the maintenance of safety was ambiguous. This ambiguity is discussed in detail later.

Uncertainty regarding future Throughout the period in which the majority of the incidents under consideration took place, there was considerable uncertainty regarding the restructuring and upgrading of the Australian ATS system. Evidence of the negative effect of this situation on performance was that some controllers were discussing possible postings and industrial issues at the time safety standards were breached. This aspect is further considered later.

Review of findings regarding psychological precursors

It was apparent from a review of the various psychological precursors of unsafe acts identified in this investigation that there was a high level of inter-relationship between them, and that many interacted to facilitate the occurrence of unsafe acts.

The evaluation also suggested some generic characteristics of controllers which had developed within the Australian ATS environment. These were an over-reliance upon individual knowledge of the workings of the ATS system and of aircraft performance, together with an excessive reliance upon aircraft actually performing as expected.

At the time of the study, it was considered that the organizational culture which had evolved in the Australian ATS system would resist change unless the proposed new air traffic management principles and training philosophies were carefully developed in concert, and reinforced with concurrent education and training programmes as well as changes within the CAA's organizational climate.

Organizational deficiencies/latent organizational failures

An important argument of the Reason model is that management must focus on identifying and rectifying those organizational deficiencies which have the greatest influence upon system safety.

This investigation revealed four distinct yet overlapping organizational deficiencies. These deficiencies were: strategic planning for air traffic management; strategic planning for training; the organizational climate; and, quality assurance. The elements upon which theses deficiencies were identified are shown in Figure 5.5.

Strategic planning for air traffic management The Australian air route structure exhibited many of the characteristics of a system which had evolved in an *ad hoc* way, rather than one which had been planned and

- Route structure made controlling more difficult
- Complex, *ad hoc* route structure   *STRATEGIC PLANNING*
- Five regional systems, not one   *FOR AIR TRAFFIC*
- Inadequate planning for traffic   *MANAGEMENT*
  changes
- Training and air traffic management did not have consistent objectives
- OJT and *ab initio* training not   *STRATEGIC PLANNING*
  coordinated   *FOR TRAINING*
- Inadequate management of OJT content and direction
- Little training to reduce human error
- Excessive reliance on controller skill   *ORGANIZATIONAL*
- Failure to effectively limit   *CLIMATE*
  controller distractions
- Ambiguity within organization
- Inadequate quality assurance to   *QUALITY ASSURANCE*
  provide ongoing feedback on key issues

Figure 5.5. Organizational deficiencies identified

developed on the basis of strategic guidelines to meet changing needs. Two major trends reflected the approach of ATS management at the time of the investigation:

- *Local fixes*: as incidents and other events revealed apparent failings in the system, fixes were applied. In many cases, such fixes were devised and implemented at a local level. Neither the broader ramifications of the problem, nor the implications of the local fix, appeared to have been fully considered on a national, system-wide basis.
- *Five systems*: at the time of the study, the CAA policy of devolution of ATS management and responsibilities from central office to the various field offices of the Authority had facilitated the separate development of five differing operational philosophies

126

within the Australian ATS system. This had meant that local solutions have not only differed in their details, but also in their underlying approach to the problems encountered.

As evidenced by the incidents under review, the lack of strategic planning had resulted in a trunk route structure which was overly complicated. In many cases, this complexity served to make the task of safely separating aircraft harder than it need have been. In the past, such complexity was accommodated by reliance on the skills of controllers. However, as traffic levels have grown, controllers have been increasingly placed in situations where they had diminishing room to manoeuvre, and were sometimes forced to adopt non standard procedures to separate the traffic.

A further result of this incremental approach to air traffic management had been a failure to plan adequately for changes in traffic patterns. This forced air traffic management into an increasingly reactive mode, especially in an environment in which the industry was demanding a more efficient and less costly service.

A more subtle ATS management issue was that the design criteria for systems and procedures did not appear to have placed sufficient emphasis on failsafe design. The inevitability of human error should be taken into account in the design of ATS systems, as it is in many areas of aviation and other industries. Many aspects of Australian airspace management did not exhibit this fundamental requirement of system safety.

Strategic planning for training   During the investigation, various aspects of the training of controllers were identified as organizational deficiencies. Considered in total, these deficiencies highlighted an overall need to regard training as a strategic issue – that is, to view the process and delivery of training as a set of national strategies aimed at achieving the CAA's objectives in the ATS arena. This matter will be further discussed in terms of corporate directions later in this chapter.

Evidence suggested that the training programme intended to develop and consolidate the skills of air traffic controllers suffered from serious deficiencies. For example:

- 'on the job training' (OJT), in terms of the quality of training which the trainee received, was not adequately managed, and standards were not established and met;
- 'inexperienced' controllers were used as OJT instructors, leading to an overall reduction in the depth of knowledge and experience passed on to the trainees;
- consolidation of skills following ratings had been hampered by a

lack of facilities, such as simulators, and by the difficulties of allocating OJT trainees quality time;
- there was a lack of formal selection and training processes for OJT trainers.

The lack of long-term strategic planning was reflected in the cyclic flow of *ab initio* trainees into the ATS system, the resources available to conduct training within the Area Approach Control Centres (AACCs), and the type of training.

Recruitment of *ab initio* trainees was out of phase with the operational requirements for controllers. In recent years, ATS management had been somewhat distanced from the *ab initio* trainee in location and organization, and had seemingly exercised little quality control over the trainees which it received.

Often, the *ab initio* trainees were initially based in AACCs where there had been little recent experience of training such individuals, and little external support for check and training officers. Evidence suggested that critical support functions such as the selection and training of trainers were inadequate, and even if resources had been available for training, there was little capacity within the system to release the training officers necessary to conduct training courses. Even rudimentary items such as training manuals were unsatisfactory, and did not meet the needs of the training officers.

The potential of training aids such as simulators also seemed to have been neglected. The simulators *in situ* at a number of the AACCs were unable to provide the complexity and variety of traffic conditions needed to properly test the performance of trainees or rated controllers.

Ratner (1987) identified problems of defensive control, decision-making and judgement. He suggested that while ATS staff were trained in the basic skills and procedures necessary for performing various duties, their training was deficient in strategic aspects. It was not providing controllers with an understanding of their fundamental capabilities and limitations with regard to human information processing.

The ongoing development of controllers, once they were part of the ATS system, seemed to have been overlooked, as the emphasis was on checking rather than training.

Organizational climate   As indicated earlier, the nature of the incidents reviewed indicated that the ATS system was unduly reliant on the skill of the controller. The system had bred, and continued to breed, a population of controllers who took considerable pride in being able to handle high traffic densities, and who were unwilling to accept modifications which might reduce the challenges.

Organizational climate may be viewed, in part, as a product of senior management. In the case of the ATS system in the period under review, there were elements of ambiguity and uncertainty within the organization. A climate existed in which devolution of responsibility to the regions had resulted in a lack of standardization at the workface. Higher management had not ensured that the process of devolution was planned and directed in a controlled and integrated manner.

In the specific period under review, ambiguity for controllers existed not only in regard to their role in the existing system, but also to their place in the future ATS system. Uncertainty regarding job positions and location, along with a lack of information regarding the introduction of the 'two centre concept', resectorization, and local management changes had led to dissatisfaction and a questioning of job worth. Uncertainty and dissatisfaction have the potential to impact on any controller's ability to operate at the highest level of efficiency and safety through distraction, subconscious influences and emotional concern.

Quality assurance   In 1991, the ATS system had the appearance of five independent systems attempting to work as one. There was a lack of standardization between these five systems. In this fragmented organizational environment, the entire Quality Assurance (QA) function within the CAA's ATS division was the responsibility of just one full time officer. The limited resources available restricted the operation of the QA function to audits of various facilities, in which the majority of the manpower was seconded to the QA function for the period of the audit (usually one week). Reviews were also conducted following serious incidents. The limited dedicated full time resources meant that the QA function was unable to:

- achieve the oversight role which was necessary for the standardization of operations and training;
- determine whether facilities met the established operating standards; and
- consider the system safety implications of projects.

Review of findings – organizational deficiencies

Reviewing the organizational deficiencies, it was apparent that the flawed organizational processes which could be identified were: poor planning in certain areas; inadequate overall monitoring of the total system; and inadequate quality assurance. It could also be argued that the ambiguity which existed between service and safety reflected an incompatibility of organizational goals.

In practice, to reduce the need for controllers to make operational decisions based purely on their knowledge of the system, which can lead to unsafe acts, it is necessary to provide an environment in which the requirements to make these kinds of decisions are limited. Such an environment can be achieved by an integrated approach to training and air traffic management which ensures that training is relevant, timely and appropriate.

Summary

This BASI airmiss study showed that the CAA exhibited organizational deficiencies similar to other large socio-technical systems charged with operating and providing a service at minimum cost to the stakeholders while maintaining high levels of safety.

The flawed organizational processes identified related mainly to planning and training. It was found that the defences of the ATS system could be substantially improved, and that there was also a need for CAA management to define explicitly the balance between the organizational goals of maximum safety and maximum, cost-effective, service.

Actions by the CAA

Although BASI has the responsibility for the independent investigation of air safety occurrences, it works closely and cooperatively with the CAA. The Bureau relies on the willingness of CAA officers to assist with investigations, and during the investigation process there is frequent contact and exchange of information, on the strict condition that no information obtained by BASI can be used by the Authority for purposes of attributing blame or liability. In this way, the CAA is often able to make changes and implement improvements well before the Bureau releases its final reports on occurrences.

Throughout this systemic investigation, CAA ATS management was invited to contribute to the project, and was given regular presentations by BASI on the progress of the analysis and findings. Consequently, by the time the investigation had been completed, the CAA had addressed many of the deficiencies identified, both in the earlier individual investigations of the individual incidents and in this systemic investigation.

Similarly, since the completion of this investigation and the second of the comprehensive reviews of the Australian ATS system carried out by Ratner (1992), which was jointly funded by BASI and the CAA, the ATS division has implemented significant changes to rectify the deficiencies identified in its organizational processess.

Training

The CAA began to implement a programme to improve the training of ATS officers. The training programme was to be reshaped so that it would:

- improve the quality of *ab initio* training;
- improve the selection system for *ab initio* trainees;
- manage their training on a three year cycle; and,
- cover the restructuring of the airspace and new airspace management.

A training manager was appointed to oversee the design of a long-term training system. A group of key ATS managers, with representatives from training organizations and the staff association, were to assist in the development of the training system. This system would include:

- skill- and competency-based training;
- refresher training and development programmes for career training; and
- the establishment of a training school in close proximity to one of the major centres (Melbourne or Brisbane), with a training annex at the other facilities.

In the shorter term, the CAA decided to introduce programmes aimed at:

- team leader and team training with the emphasis on leadership, team building, staff development, standards, performance assessment, care of staff and productivity;
- enhancing the quality of instruction;
- developing cost-effective use of computer-based training systems;
- more comprehensive and structured use of simulators to provide more effective training before entering the OJT system;
- building ATC's awareness of human performance capabilities, their limitations and other Human Factors in the operational environment; and
- a training course in ATS incident investigation has been introduced with assistance from BASI and the Airways

131

Corporation of New Zealand. The Reason model provides a central framework for this course.

Other initiatives taken by the CAA included the participation of air traffic controllers in Crew Resource Management courses conducted by the airlines, and negotiations with the Royal Australian Air Force (RAAF) on the use of its military control tower visual simulator training facility.

Standard procedures and traffic management

The CAA sent a team on a two week fact finding visit to the United States to examine Standard Terminal Arrival Routes (STARS). The terms of reference for the project included:

- the establishment of a draft policy for the use of STARS within Australia;
- the criteria for the design, implementation and publication of STARS for use at major capital city airports;
- the design of terminal area and en route procedures to maximize the efficiencies to be gained from STARS; and
- the identification of coincident changes needed for the implementation of STARS.

The CAA team recommended to the Authority that STARS be implemented whenever efficiencies in aircraft operations, traffic management and ATS workload/coordination can be gained. The relevant US policy was to be used as a guide.

The team also recommended that the requirement to include controlled airspace boundaries in the design of Standard Instrument Departures (SIDs) be reviewed. Other recommendations involved traffic management, procedures, standards, publications, control tower operations and equipment.

Many of these recommendations addressed areas identified in the previous BASI incident investigations, the present airmiss study and the Ratner review.

Quality assurance (QA)

The CAA has developed the ATS QA unit by expanding its role and staffing. There is recognition and senior management commitment that the QA Unit should play a vital part in ATS management. The main impetus for this change in the QA Unit came from the 1991

Ratner review of the ATS System. Senior management is publicly committed to the implementation of all of the Ratner Review's recommendations, which primarily addressed systemic organizational process issues. The substantially enhanced ATS QA Unit has played an important role in this process.

An integrated strategic plan for ongoing quality assurance reviews of the CAA's ATS facilities has been developed by the QA unit. These reviews commenced in 1993 by teams under the direction of the QA officers, and included participation by other officers from agencies such as BASI, the Royal Australian Air Force and the Airways Corporation of New Zealand.

Management

The CAA has changed the ATS management structure to align with the changes in the ATS regional sub-divisions. The Assistant General Managers (AGMs) have been moved closer to the regional ATS facilities where practicable. Similarly, where possible, their officers have been relocated to the airport complexes to ensure better feedback and awareness of issues at the workface.

Safety actions and recommendations

As outlined in the previous section, since the commencement of the airmiss study, CAA ATS management initiatives have addressed some of the organizational factors identified. Other initiatives will be taken as resources and opportunities arise. However, the Bureau believed that a number of further initiatives were required. These are presented below.

Strategic planning

It was apparent that there was a need for a coordinated integrated systems approach to the planning of air traffic services, involving the regulatory division of the authority as well as the ATS division. This approach would involve assessing the risks associated with any changes during planning, monitoring of the implementation of special projects, providing appropriate feedback once the initiatives have been implemented, and ensuring that the full potential of each new development is realized. BASI suggested that this should be the role of a dedicated strategic planning unit.

It was the Bureau's contention that the CAA should take a proactive approach to identify, and continuously monitor, areas of weakness within the ATS system. As stated in Reason's analysis of British Rail:

> negative outcome data (accidents, incidents, etc.) are too sparse, too late and too statistically unreliable to support effective safety management.

More valid indicators of an organization's safety health are required, by which may be measured its intrinsic resistance to combinations of weakened or breached defences, unsafe acts, technical failures and unusual environmental effects acting on the system. BASI suggested that the CAA has such a requirement, that is a structured approach which can monitor the vital signs of the health of the organization, while at the same time also providing feedback on the success of new initiatives. A 'system safety' group, which operates across both ATS and the regulatory division, should be established to fulfil this requirement.

Following reviews by Reason and his colleagues, British Rail, Shell and British Airways have introduced innovative programmes to provide measures of system safety which can be applied routinely in a proactive mode, and intermittently in a reactive mode (i.e. following an accident or incident; see Chapter 6). The indices utilized include planning, communication and training procedures. The adequacy of the current state of each index, and the potential for safety problems, are continuously monitored and assessed by the organization.

BASI stated that it would be willing to provide assistance with the development and implementation of a proactive systems approach, whereby the CAA:

1. Coordinates and integrates the planning and implementation of special projects and evaluates the success of the projects once the initiatives have been implemented.
2. Evaluates the potential of such an approach, which attempts to provide valid ongoing indicators of the organization's safety health (e.g. PRISM, MESH, TRIPOD) in aiding the QA function.

Corporate directions

Although, as noted earlier, much has been done, the Bureau believed that in this period of major organizational and technical change, there remained a need for further initiatives to promote the safety health of the organization. The Bureau suggested that the Civil Aviation Authority consider adopting a similar safety philosophy to that utilized by the United Kingdom's Civil Aviation Authority, which was highlighted in the 1992 Ratner Review.

Following the airmiss study, the Bureau has continued to work closely with the CAA's expanded ATS QA unit. A series of regular meetings was initiated after the joint Ratner review, to ensure better communication and liaison between the two organizations. These meetings have proved very effective, and it was through this channel that the progress reports and results of this systemic investigation were promulgated.

Safety net

Given the fundamental capabilities and limitations of human performance, the Bureau recommended that the CAA:

1. Undertake an active assessment of ground and airborne technologies for collision prevention.
2. Continue to optimize controller workload by standardization of procedures and resectorization, with the aim of reducing the number of unsafe acts.
3. In conjunction with the Australasian Airlines Flight Safety Council and the Australian Aviation Industry Association, promulgate the role which aircrew can play in the detection of controller error.

Conclusion

This airmiss investigation reflected the new proactive and systemic approach of the Bureau, and was based on the structured approach developed by Reason for which the International Civil Aviation Organization (ICAO) has since issued guidelines (ICAO, 1993, 1994). The investigation examined active and latent failures of the ATS system which were associated with the airmiss occurrences. These unsafe acts (active failures) were categorized as errors or violations. In addition, psychological precursors of unsafe acts, organizational deficiencies and corporate actions were also identified (latent failures).

Deficiencies in the CAA's strategic planning of the air traffic management and training were found to be major factors in these active and latent failures. The formation and operation of an effective Quality Assurance function was highlighted as one of the main potential improvements to the ATS system.

A number of major initiatives have been introduced by the CAA since the completion of this systemic investigation. These initiatives have addressed many of the problems. However, there remain some organizational aspects which still require attention. Specifically, the Bureau has suggested that the CAA should:

- introduce further system safety initiatives based on the MESH, TRIPOD and PRISM models used by organizations such as British Airways;
- adopt a similar safety philosophy to that used by the Civil Aviation Authority in the United Kingdom; and
- undertake an assessment of ground and airborne technology for collision prevention.

The report also emphasized that aircrew play a vital role in the 'defensive' mechanisms of the ATS System. The Bureau therefore recommends that the CAA explores ways of enhancing the aircrew's detection of errors in the controller's actions.

Postscript

In August 1994, BASI released a report of its investigation into a fatal commuter airline accident (BASI, 1994b). Like the airmiss report described in this chapter, this accident report was structured in accordance with the Reason model.

In its response to the accident report the CAA announced that it was embracing the systemic approach to aviation safety management, and that, in line with this initiative, it was establishing an Aviation System Safety Branch (CAA, 1994).

Acknowledgements

As with most BASI reports, this airmiss study was a team effort by a number of specialist staff. Primary members of the team were Bob Dodd, formerly exchange officer to BASI from the Canadian Transportation Safety Board and newly appointed manager of the Aviation System Safety Branch of the CAA, Peter Reardon, ATS investigator, Dr Claire Marrisson, human performance investigator, and Paul Mayes, Manager Safety Analysis. In addition, BASI line investigators made significant contributions to the project.

References

BASI (1994a), *A Systemic investigation of Airmiss Occurrences 1991–1992.* Report RP/91/11, BASI Canberra.

BASI (1994b), *Piper PA31-350 Chieftain, Young NSW, 11 June 1993.* Investigation Report 9301743, BASI Canberra.

CAA (1994) Authority adapts system approach to safety. *Aviation Bulletin*, Vol. III, No. 7, pp 1ff. CAA, Canberra.

Charlton, R. (1993), Integrating work and safety: a personal view, *15th Annual Colliery Safety Symposium*, Sydney, Shell Australia, Melbourne.

Freeman, C. and Simmon, D.A. (1990) Crew Resource Management model, *Proceedings Flight Safety Foundation 43rd Annual International Seminar*, Rome, FSF, Arlington.

ICAO (1993) Investigation of Human Factors in accidents and incidents, *Human Factors Digest No. 7*, Circular 240-AN/144, ICAO, Montreal.

ICAO (1994) Human Factors, Management and Organisation, *Human Factors Digest, No. 10*, Circular 247-AN/148 ICAO, Montreal.

ICAO (1970) *Manual of Aircraft Accident Investigation*, 4th Edition, ICAO, Montreal.

NTSB (1991) *Runway Collision of USAIR Flight 1493, Boeing 737 and Skywest Flight 5569 Fairchild Metroliner, Los Angeles International Airport, Los Angeles, California*, NTSB/AAR-91/08, NTSB, Washington.

O'Hare D. (1994), Latent failure and human error. In *New Zealand Flight Safety, FSM-94-1*, Civil Aviation Authority, Lower Hutt.

Ratner, R. (1987), *Review of the Air Traffic Services System of Australia*, Dept of Aviation, Canberra.

Ratner, R. (1992), *1992 Review of the Australian Air Traffic Services System*, Civil Aviation Authority, Canberra.

Reason, J. (1993) Organisations, corporate culture and risk. In: *Human Factors in Aviation, Proceedings of the 22nd International Air Transport Association's Technical Conference*, IATA, Montreal.

6 Remedial implications: some practical applications of the theory

Introduction

Throughout this book, we have stressed the importance for aviation safety of going beyond the individual 'human factor' to embrace the task, situational and organizational factors that shape 'sharp end' performance. It is our contention that the purely psychological influences upon the reliability of skilled behaviour (e.g. preoccupation, distraction, misperception, momentary forgetting, and the like) are the last and least manageable elements in the sequence of events contributing to the errors of pilots, air traffic controllers and maintenance engineers.

In the past, the probable or established proximal failures of those at the immediate human–system interface have tended to dominate the concerns of accident investigators and regulators. But, as we have argued, unsafe acts on the front line do not constitute an adequate cause of accidents, even though such categories as 'pilot error' are enshrined in the history of air accident investigation. In an industry where the standards of selection, training and the monitoring of technical skills are among the most rigorous to be found anywhere, the rare instances of catastrophic error or procedural violation are more likely to be the *consequences* of pre-existing systemic failings rather than the sole or sufficient causes of an accident. And even on the rare occasions where this cannot easily be demonstrated, it still makes more remedial sense to opt for a latent rather an active failure view of accident causation, since the former is more easily managed than the unpredictable and ephemeral mental states that can be the immediate precursors of an error.

As mentioned in the first chapter, a theory of accident causation is only as good as the effectiveness of its practical applications. The aim of this final chapter is to outline some of the remedial implications of the views expressed throughout this book. Such practical measures fall into two mutually dependent groups: *reactive measures* that can only be applied *after* the occurrence of an incident or accident, and *proactive measures* that can be implemented *before* the occurrence of such events to assess the 'safety health' of workplaces and organizations. Effective safety management requires the application of both of these measures. We need to learn the right lessons from the accidents and near misses that will always occur in hazardous systems. And we also need to make visible to those who manage and operate the system the latent failures and resident pathogens that are an inevitable part of any complex technology, no matter how well run, and which experience has shown can set up the conditions likely to provoke future accidents.

Why reactive and proactive measures are both necessary and mutually interdependent

The accident theory presented in this book directs us to focus our remedial efforts upon two major issues:

- The task, situational and systemic factors that, in combination, are known to promote errors and violations. These will be referred to collectively as *error-promoting conditions* (EPCs), where, for the sake of brevity, the term error includes violations as well as slips, lapses and mistakes (see Chapter 1). The causal argument goes as follows: strategic decisions taken in the upper echelons of the organization create inevitable latent error-producing conditions within specific workplaces; these then interact with mental and physical states to generate unsafe acts.
- The barriers and safeguards that mitigate or block the adverse consequences of errors and violations. These will be referred to collectively as *defences*, and may be classified (see Chapter 1) according to the functions they serve (i.e. alarm, warning, protection, recovery, containment, escape and rescue) and by the various systemic modes in which they are implemented (i.e. engineered safety features, supervision, safety briefings, administrative controls, rules and regulations, personal protective equipment, etc.).

The reactive and proactive applications outlined above each play mutually supportive roles in regard to identifying both the presence and significance of error-promoting conditions and the existence of defensive weaknesses or absences. These roles are described below.

Error-promoting conditions

Reactive applications There are potentially a large number of EPCs. The more significant of these for various aviation contexts can be revealed by analyses of a number of domain-specific accident investigations – assuming, of course, that these investigations took adequate note of the situational and organizational factors implicated in the event. Analyses over several domain-related accidents can reveal patterns of cause and effect rarely evident in single-case investigations.

Procative applications The 'safety health' of a workplace or organization, like that of an individual, cannot be derived directly from a single measure; it must be inferred from the values obtained on a variety of 'vital signs'. Appropriate after-the-event analyses can identify which of the many possible indices are likely to yield the most diagnostic assessments in any given context (i.e. flight deck, air traffic control centre, maintenance hangar, etc.). Once identified, these significant indices can be measured at regular intervals to provide: (a) an overall indication of current 'safety health' – where 'safety health' means the extent to which the system is able to resist or withstand the ever-present accident-producing factors encountered during its normal course of operations; and (b) an identification of those situational and organizational factors that are most in need of correction at that particular time. The frequency of these assessments depends upon the rate at which the process in question is likely to change. Local factors change faster than organizational factors, and so need to be monitored at more frequent intervals.

Defences

Reactive applications Each closely investigated event documents a partial or complete accident trajectory through a system's 'defences-in-depth'. These snapshots are important because they reveal the often diabolical and improbable ways in which Sod, Murphy and malign chance can string together human, technical and system failures to breach even the most sophisticated defences and bring hazards in damaging contact with victims and assets. Once again, analyses of several incidents can reveal recurrent and often hard-to-imagine patterns of defensive weaknesses.

Proactive applications Regular assessments of the integrity and effectiveness of the various defences tell us where the 'holes' in each layer (see Chapter 1, Figure 1.8) presently exist, or are likely to appear in the future. There is no such thing as a perfect defensive arrangement, but regular checks upon the system as a whole will lead to a continuous improvement cycle in which the worst problems identified on any one assessment are corrected, with other pressing problems being brought to light on subsequent test occasions, and so on. In this way, safety management changes from being a process dominated by retrospective repairs (i.e. fixing the stable door after the horse has bolted) to one in which prospective reform plays the leading part (i.e. making a stable from which no horse could run, or even want to).

The remainder of this chapter will examine some reactive and proactive measures derived from the accident causation theory. Some of these are relatively well established, others are still in the research and development stages. In each section, we will first outline the general principles involved in the application (either reactive or proactive), and then flesh these out with practical aviation examples.

Reactive applications of the theory

The practical challenge can be expressed quite simply: how can we develop an accident investigation tool that will lead its users in a step-by-step fashion back to the kinds of organizational and managerial root causes identified by the Dryden (Moshansky, 1992) and Strasbourg (Pariès, 1994) inquiries? And there is an important rider: how can we do this without incurring vast investments in time, money and personnel? Clearly, whatever tool is devised has to be appropriate to the limited resources available to most accident investigation agencies.

A research team at the University of Manchester was recently commissioned to create an accident investigation tool suitable for investigating accidents and incidents on British railways. The tool was call RAIT (Railway Accident Investigation Tool), and was initially applied to infrastructure accidents (i.e. accidents to shunters and permanent way workers). Although devised for a railway rather than an aviation domain, its general principles are applicable across a wide range of transport activities, as well as to other complex, well-defended technologies. These general principles are summarized below.

141

Organizational and managerial factors

Ten Railway Problem Factors (RPFs) were selected as being representative of significant organizational and managerial root causes of railway infrastructure accidents. There are listed below. A detailed description of their sub-components is given elsewhere (Reason, 1993a, b):

- Training
- Provision of tools and equipment
- Materials
- Design
- Communications
- Rules
- Supervision
- Planning
- Commercial and operational pressures
- Maintenance

The main components of the model have been described in Chapter 1. The model identifies two causal pathways linking these organizational processes with the occurrence of accidents and incidents:

- *Active failure pathway.* This pathway begins in the managerial and organizational spheres with decisions regarding the ten RPFs. All such decisions, even good ones, are likely to carry a downside for someone at some point in the system. These negative consequences (organizational pathogens) are manifested within the various workplaces as error-promoting and violation-enforcing conditions (e.g. time pressure, unawareness of hazards, etc.). In performing a particular task, these local factors can interact with personal factors (i.e. cognitive or motivational problems) to create errors and violations. Some of these may act to penetrate or bypass one or more of the defences, leading to an accident or incident.
- *Latent failure pathway.* This pathway reflects the ways in which organizational processes can have adverse effects upon defences without the intervention of unsafe acts (active failures) in the workplace. The nature and variety of these defensive weaknesses are discussed further below (see Step 2).

RAIT: *the analytical steps*

RAIT is an implementation (operationalization) of the accident causation model as a practical investigative tool. It begins with the outcome and works back to the organizational root causes along both the active and the latent failure pathways. It can operate in either an *investigative mode*, when it is employed by accident investigators to determine the root causes of a given event, or in an *analytical mode*, when it is applied to a number of domain-specific accidents in order to determine recurrent patterns of cause and effect.

For each accident (in either mode) there are six main steps. These are summarized diagrammatically in Figure 6.1, and expanded below:

Step 1: Describe the outcome: fatality, injury or near miss.

Step 2: Identify the latent weaknesses in defences and barriers. This was done using a matrix of defensive modes by functions served. The defensive modes were: supervision (relative to safety issues), defects reporting, engineered safety features, safety briefing/training, certification, Rule Book, etc., personal protective equipment, protection systems, standards, procedures and administrative controls, hazard identification and elimination. The defensive functions served by these modes were: creating awareness and understanding of hazards, detection, warning, protection, recovery, containment, escape and rescue. Pre-existing defensive failures (i.e. latent failures) together with on-the-spot attempts to bypass or circumvent existing defences (i.e. active failures) were classified by placing them within the appropriate cells of the matrix.

Step 3: Identify the nature of the unsafe acts. The available data only permitted a three-way classification: errors, violations and erroneous violations.

Step 4: Identify and categorize the tasks or activities associated with these unsafe acts.

Step 5: Identify (for the accident as a whole) local factors likely to have provoked or promoted the unsafe acts.

Step 6: Determine the organizational root causes of the accident. This involved three distinct stages:

> (a) Each of the latent defence failures, identified in Step 2, was assessed for its relative contributions from each of the ten Railway Problem Factors (RPFs), using 0–4

143

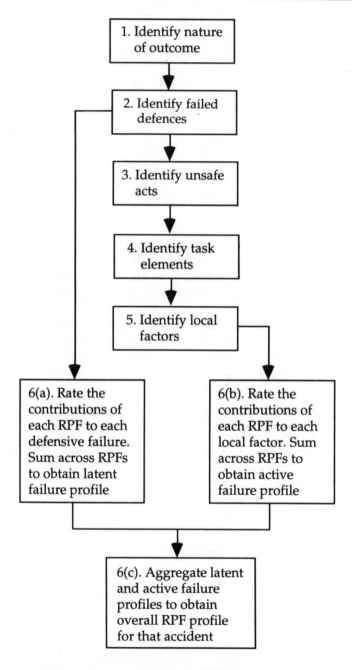

Figure 6.1. Summarizing the steps involved in carrying out a RAIT analysis on a particular accident

144

ratings. When these ratings were summed over latent factors, they yielded a latent failure (organizational) profile (i.e. a visual indication of the relative contribution of each organizational factor (OF) to the latent factors identified in that accident).

(b) The same basic procedure was repeated for each of the local factors identified in Step 5. This yielded an active failure (organizational) profile, showing how each of the RPFs contributed to the local factors for this accident as a whole.

(c) The latent and active profiles were aggregated to give an overall organizational profile for that accident. This identified those RPFs most responsible for both the latent and the active failure pathways.

Results from RAIT analyses of 57 infrastructure accidents

When applied to the inquiry reports of 57 railway infrastructure accidents (i.e. in its analytical mode), RAIT showed itself to be a reliable accident investigation tool that is capable of identifying patterns of managerial, organizational, technical and operational root causes in a cost-effective fashion. The study yielded the following general findings:

(a) RPF profiles (latent, active and aggregate) showed a high degree of consistency regardless of the analyst, the accident outcome and the method of analysis.

(b) When all 57 accidents were considered collectively, four RPFs appeared to dominate the risks: supervision, communication, training and planning. In a number of acidents, however, more localized technical and operational factors (e.g. tools and equipment, materials, maintenance, etc.) played a significant and sometimes dominant role in the development of the accident sequence.

(c) Analyses of defensive weaknesses highlighted the problems associated with creating adequate awareness and understanding of the hazards. The greatest deficiencies here were in the area of supervision and briefings.

Problems with RAIT

Despite these generally satisfactory findings, a number of problems arose in applying the tool in this way. The first was that by working through the noisy medium of accident reports, the RAIT researchers were dealing with second- or third-hand information that was already pre-shaped by the interests and biases of the original investigators. Many of the questions to which answers were needed had not been asked by the investigators. Clearly, many of these problems would not apply when the tool is applied directly by scene-of-accident investigators.

A second problem related to judging the contributions of the ten RPFs – the organizational and managerial root causes. The RPFs divided into two clusters. One group (i.e. training, supervision, planning, communication and rules) related primarily to general issues; the other five were more site-specific. Overall, the five general RPFs accounted for 80% of the causal factors.

It could be argued that this result reflects *generality differences* in the RPFs. Management and organizational issues, like supervision, communication, planning, training and rules, are more pervasive in their influence (and are *seen* to be more pervasive) than are site-specific factors such as design, the provision of tools and equipment, maintenance and materials. As a result, they could have attracted both more causal attributions and higher ratings from the RAIT analysts. In addition, there is often insufficient information provided by the accident reports to *exclude* the possible influence of these more general factors.

An analogy will help to clarify the matter. Imagine that the ten RPFs are dartboards, and that the causal attributions are darts. If all the targets were the same size and the analysts were throwing their darts 'blind', one would expect, by chance, a fairly equal number of hits across the ten 'boards'. But that is not what was found: there were obviously large differences in 'hits'. But suppose the dartboards were of different sizes – as suggested above – then, by chance, the larger boards would attract more 'hits' simply by virtue of their size.

This, of course, is the harshest possible interpretation: namely, that the results were entirely due to a combination of perceived differences in generality and chance. In reality, the researchers were not making their attributions 'blind', but they were working through the distorting medium of the accident reports and at considerable remove from the realities of the events. It is also possible that the original investigators, wittingly or unwittingly, highlighted the more general managerial issues at the expense of the local factors. If this were the case, the analysts could merely have been reflecting the biases of the investigators.

However, the fact that some accidents were clearly associated with constellations of the more local RPF contributions weakens the potential criticism that the consistent RPF profiles (showing the dominance of managerial and organizational factors) were due solely to differences in *generality* between the RPFs. But there is a strong case for carrying out RAIT-like analyses using more than the ten problem factors included here. It is believed that the additional effort and time involved in carrying out these fuller analyses would be outweighed by being able to distinguish, for example, between staff and departmental communications. It would also be advisable to consider ways of breaking down the supervisory and training RPFs into more specific and more easily targeted sub-categories.

Application to other domains

What has been described above is an application of the model to a highly specific domain, the railway infrastructure. Although the basic steps could still be used for air accident investigations, it is likely that organizational and managerial factors other than the ten RPFs employed by RAIT would be required, though there is likely to be a large degree of overlap between the RPFs and more aviation-specific systemic factors. An example of a RAIT analysis in aviation is described below.

RAIT-like analysis of Controlled Flight Into Terrain (CFIT) accidents

Controlled flight into terrain accidents can be defined, in broad terms, as those accidents in which a perfectly sound aircraft, without any defects or malfunctions, is inadvertently flown into terrain (or water), without previous awareness by part of the flight crew of the impending disaster. The DC10 crash into Mount Erebus (see Chapter 2) is a classic example of a CFIT accident. According to widely-publicized statistics, CFIT accidents account for the vast majority of aircraft accidents over the last years. There is widespread consensus of opinion among the international aviation community that urgent remedial measures are needed to contain these accidents.

In late 1994, the International Civil Aviation Organization (ICAO) undertook, within the context of its Flight Safety and Human Factors Programme, an analysis of 24 selected CFIT accidents. The analysis followed both the methodology and philosophy advocated in this book. The tool employed to conduct the analysis was the general accident causation model discussed in Chapter 1. The accidents selected were considered to constitute a representative cross-sample of different kinds of operations and environments in different parts of the world. The list of accidents analysed is included in Table 6.1.

147

Table 6.1.   List of CFIT accidents

Aircraft Occurrence Report, Nahanni Air Services Ltd. de Havilland of Canada DHC-6-100 C-FPPL, Fort Franklin, Northwest Territories, 9 October 1984. Report Number 84-H40004

Aviation Occurrence Report, Labrador Airways Ltd. de Havilland of Canada DHC-6-100 C-FAUS, Goose Bay, Labrador, Newfoundland, 11 October 1984. Report Number 84-H40005

Aviation Occurrence Report, Simpson Air Ltd., Beechcraft King Air B-90 C-GDOM, Fort Simpson Airport, Northwest Territories, 16 October 1988. Report Number A 88W0234

Aircraft Accident Report, Embraer 110 Bandeirante, OH-EBA, in the vicinity of Ilmajoki Airport, Finland, November 14, 1988. Major Accident Report No. 2/1988, Helsinki, 1990. Ministry of Justice, Ilmajoki Aircraft Accident Investigation Board

Aviation Occurrence Report, Voyageur Airways Ltd. Beechcraft King Air A-100 C-GJUL, Chapleau, Ontario, 29 November 1988. Report Number 8800491

Aviation Occurrence Report, Air Creebec Inc., Hawker Siddeley HS 748-2A C-GQSV, Waskaganish, Quebec, 3 December 1988. Report Number A88Q0334

Aircraft Accident Report, Fairchild Swearingen Merlin III SA226T, N26RT, Helsinki-Vantaa Airport, Finland, February 23, 1989. Accident Report No. 1/1989, Helsinki, 1989. Ministry of Justice, Planning Commission for the Investigation of Major Accidents

Aviation Occurrence Brief, Ptarmigan Airways Ltd., Piper PA-31T Cheyenne C-GAMJ, Hall Beach, Northwest Territories, 17 April 1989. Brief Number A89C0069

Aviation Occurrence Report, Skylink Airlines Ltd., Fairchild Aircraft Corporation SA227 Metro III C-GSLB, Terrace Airport, British Columbia, 26 September 1989. Report Number 89H0007

Aircraft Accident Report, Aloha Islandair, Inc.. Flight 1712, de Havilland Twin Otter, DHC-6-300, N707PV, Halawa Point, Molokai, Hawaii, October 28, 1989

Report on the Accident to Indian Airlines Airbus A-320 Aircraft VT-EPN on 14th February, 1990 at Bangalore. Government of India, Ministry of Civil Aviation.

Aviation Occurrence Report, Frontier Air Ltd., Beechcraft C99 Airliner C-GFAW, Moonsonee, Ontario, 30 April 1990. Report Number A90H0002

Accident Investigation Report, Beech King Air E90 VH-LFH, Wondai, Queensland, 26 July 1990. BASI Report B/901/1047

Final report of the Federal Aircraft Accidents Inquiry Board concerning the Accident of the aircraft DC-9-32, ALITALIA. Flight No. AZ404, I-ATJA on the Stadlerberg, Weiach, 14 November 1990

Aircraft Accident/Incident Summary Report, Controlled Flight Into Terrain, Bruno's Inc., Beechjet, N25BR, Rome, Georgia, December 11, 1991

Aircraft Accident/Incident Summary Report, Business Express, Inc., Beechcraft 1900C N811BE, near Block Island, Rhode Island, December 28 1991

Rapport de la Commission d'Enquête sur l'Accident Survenu le 20 Janvier 1992 près du Mont Saint Odile (Bas Rhin) à l'Airbus A320 Immatricule F-GGED Exploite par la Compagnie Air Inter

Rapport Final d'enquête d'Accident d'Avion, CV 640, N862FW, Gamcrest-Gambie, 09 fevier 1992, Kafountine, Senegal. Ministère de l'Equipment des Transports et de la Mer. Republique du Senegal

Aircraft Accident Report, Controlled Collision with Terrain, GP Express Airlines, Inc., Flight 861 a Beechcraft C99, N118GP, June 8, 1992 – PB93-910403 NTSB/AAR-93-03

Report on the Accident of Thai Airways International A310, Flight TG 311 (HS-TID) on 31 July 1992. His Majesty's Government of Nepal, June 1993

Report on the Accident of Pakistan International Airlines A300, Flight PK268 (AP-BCP) on 28 September 1992. His Majesty's Government of Nepal, 1993

Aircraft Accident/Incident Summary Report, Controlled Flight Into Terrain, GP Express Airlines, Inc, N115GP Beechcraft C-99, Shelton, Nebraska, April 28, 1993. PB94-910401 NTSB/AAR-94/01/SUM

Aircraft Accident Report, Controlled Flight into Terrain, Federal Aviation Administration Beech Super King Air 300/F, N82, Front Royal, Virginia, October 26, 1993. PB94-910405 NTSB/AAR-94-03

Aircraft Accident Report, Controlled Collision with Terrain, Express II Airlines, Inc./Northwest Airlink Flight 5719, Jetstream BA-3100, N334PX, Hibbing, Minnesota, December 1, 1993

148

The accidents involved commercial air transport turboprop/ turbojet aircraft accidents investigated between 1984 and 1994, independent of aircraft mass or seating capacity. The data obtained from the study are factual data extracted from the official investigation reports, *without inferences or assumptions* by the analysts.

The purpose of the study was to determine whether there is a set of human performance issues as well as of organizational factors involved in CFIT accidents which consistently emerge from official reports. In other words, this study attempted to define the 'anatomy' of CFIT accidents from the perspective of both human and organizational factors.

ICAO has long recognized that operational personnel performance errors do not take place in a social vacuum, but within contexts which either resist or foster them. Although lapses in human performance were cited in all CFIT reports analysed, all of them also disclosed flaws and deficiencies of the aviation system which may adversely have affected human performance. The analysis was limited to those deficiencies in the particular aviation system identified by the official reports, *without inferences or assumptions* by the analysts.

The analysis thus presents a dual pathway leading to CFIT accidents: an 'active' pathway, generated by actions or inactions of front-line operational personnel (i.e. pilots, controllers, and so forth); and a 'latent' pathway, generated by deficiencies in various aspects of the aviation system.

The data obtained from the analysis are presented in the following paragraphs in tabular format. This information is intended to allow organizations in aviation to concentrate remedial action on the four or five most significant factors identified under each category in order to contain the scourge of CFIT accidents.

Figure 6.2 shows the fundamental importance of clear organizational policies which unambiguously indicate to front-line personnel what are the operational behaviours management expects from them in practice. It also illustrates the consequences of the conflict in the allocation of resources between production and safety goals, usually a byproduct of flawed organizational policy-making. Likewise, Figure 6.2 supports the argument (see Chapter 2) about management's responsibility in establishing structures for anticipating, containing and controlling unexpected and unanticipated events which may threaten system safety through adequate feedback regarding the quality of organizational processes.

Figure 6.3 illustrates which defences, barriers and safeguards in the aviation system are most vulnerable to flawed organizational processes when considering CFIT accidents. The consistency of the data is

149

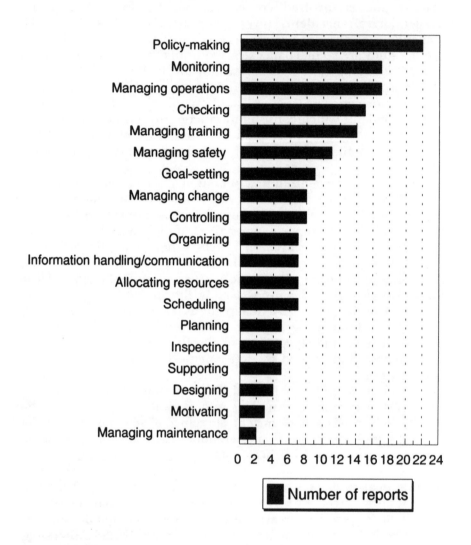

Figure 6.2. Organizational processes (based on 24 reports)

apparent, since establishing policies and standards, defining procedures and supervising their proper implementation and execution, and training (including crew coordination training) fall as a cluster under the five topmost organizational processes identified in the accident reports and included in Figure 6.2. It may be worth

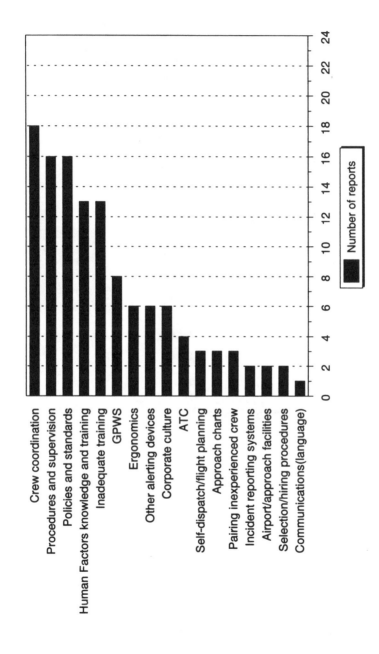

Figure 6.3.   Defences, barriers and safeguards (based on 24 reports)

151

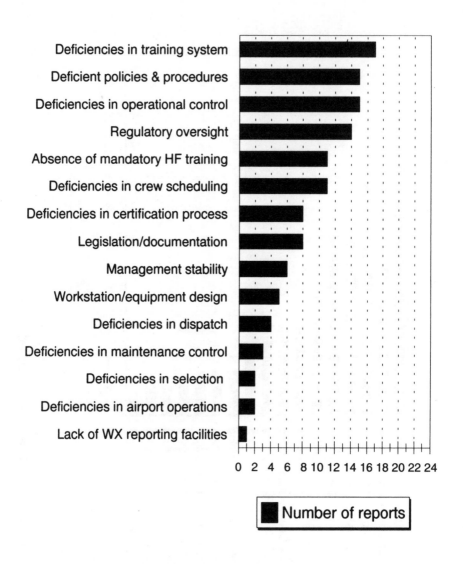

Figure 6.4.   Latent failures (based on 24 reports)

mentioning, as a side remark, that four of the six accidents in which less than optimum ergonomic design was identified as a defence which did not perform as expected involved late technology, automated aircraft. This may well constitute a word of caution against excessive reliance on technology to contain the adverse consequences of human

152

error at the expense of the broader systemic considerations.

Figure 6.4 completes the 'latent' pathway to CFIT accidents, and the data, as indicated by the three topmost types of latent failures identified, are consistent with the most significant flawed organizational

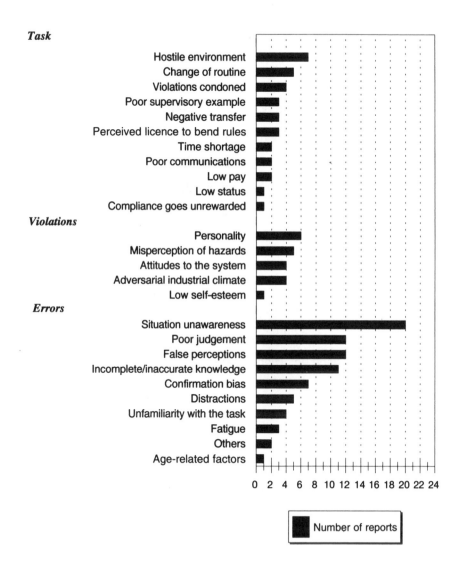

Figure 6.5.   Local working conditions (based on 24 reports)

153

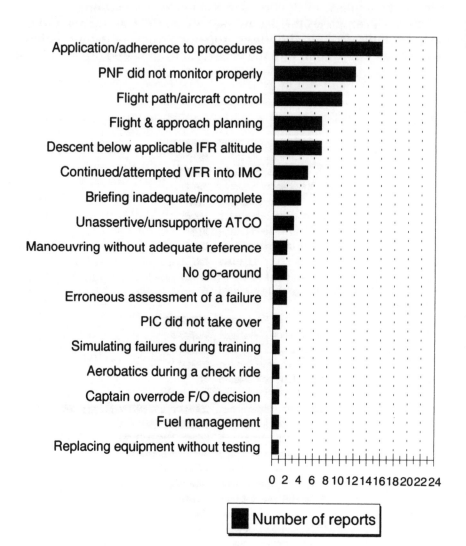

Figure 6.6.   Active failures (based on 24 reports)

processes and the most frequently broken defences identified in Figures 6.2 and 6.3. That regulatory oversight is the fourth most significant latent failure identified in the accident reports suggests – in contrast with conventional knowledge – that airlines are not the only and exclusive organizations in aviation the activities of which carry

154

potential to induce safety breakdowns and generate accidents.

Figure 6.5 shows the start of the 'active' pathway to CFIT accidents, illustrating the working conditions which detracted from the efficiency and reliability of human performance in the particular contexts within which the accidents occurred. The adverse consequences of a hostile environment can most effectively be mitigated by clear policies and by operational control and feedback. To cope with changes in operational routines, training systems must adequately address the components of change. The link between the significant task factors leading to CFIT accidents and the significant flawed organizational processes, broken defences and latent failures in the 'latent' pathway would appear reasonably obvious. On the other hand, the predominance of errors over violations would support the contention that operational personnel are, for the most part, well-intentioned individuals, trying to discharge their responsibilities to the best of their professional abilities. They succumb, from time to time, to inherent human limitations. This strengthens the argument in Chapter 1 and elsewhere in the book that exhortations and admonitions 'to do better' or 'to know better' have minimum prevention and remedial value.

Figure 6.6 identifies the non-application or non-adherence to procedures, including the failure of the Pilot-Not-Flying (PNF) to monitor the Pilot-Flying (PF) as the most significant active failures. It is essential, however, to resist the temptation to consider these errors as isolated events. In fact, when considering them within the context of the surrounding events, a pattern emerges which indicates that in most accidents analysed, the adoption of 'operational shortcuts' was too often the best way – and in many cases the only way – flight crews had to accomplish their tasks.

When summarizing the data in terms of the relative contribution of each category within the accident causation model utilized to the total sum of CFIT accidents analysed, the picture shown in Figure 6.7 emerges.

Figure 6.7 supports the assertion in this chapter – and the fundamental proposal of this book – about the need of going beyond the 'individual human factor', the relative contribution of which is in the order of 12%, into the task, situational and organizational factors that shape human performance. The remedial implications – in terms of both proactive and reactive safety management – of the data presented in Figure 6.7 are clear.

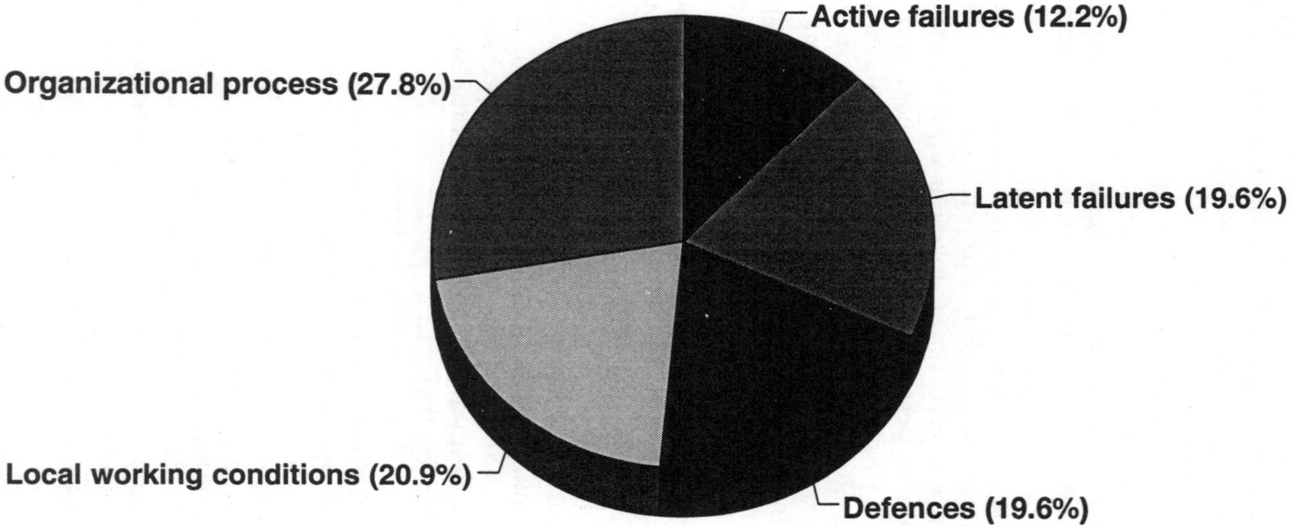

Active failures (12.2%)

Latent failures (19.6%)

Defences (19.6%)

Organizational process (27.8%)

Local working conditions (20.9%)

Figure 6.7. CFIT accidents (based on 24 reports)

156

Proactive applications of the theory

While reactive measures are an essential part of any safety management system, they suffer a number of limitations. By their nature, they reveal problems after the event. The information they provide is both too little and too late to support effective safety management. This is particularly the case in aviation, where the number of accidents is small relative to exposure, and has remained asymptotic at round 1 fatality per 10^6 flying hours for the past 20 years or so. Thus, negative outcome data are sparse, often conveying more noise than signal in their minor fluctuations from year to year. Near miss reporting systems (e.g. British Airways Confidential Human Factors Reporting Programme, see O'Leary and Fisher, 1993) help to fill this data gap and provide valuable insights into underlying organizational and situational factors, but even such excellent schemes as these are insufficient to gauge an organization's 'safety health' by themselves. They share with all other negative (or near negative) outcome data the problem that they are too susceptible to the happy and unhappy influences of chance. Good organizations can have bad accidents and even large numbers of near misses – this is not as paradoxical as it first sounds, since only good organizations are likely to receive such reports in the first place. Conversely, good luck can keep bad organizations free of accidents and incidents for long periods of time.

The two faces of safety

We need to recognize that safety, like health, has two faces. There is a negative face manifested by accident and near miss reporting systems. These reveal moments of vulnerability. Then there is a positive face, having to do with the system's *intrinsic resistance* to the accident-producing 'slings and arrows' associated with its normal activities. Whereas negative outcome data readily convert into numbers, charts and trends, postive safety 'health' is much harder to pin down and measure. To understand how health can be measured, we need to introduce the idea of the 'safety space'.

The 'safety space'

Let us elaborate a little further on these two faces of safety. The great French mathematician, Poisson, derived his well-known distribution from data such as the number of horse-kicks sustained by cavalrymen over a given period of time. The largest proportion of any unit would

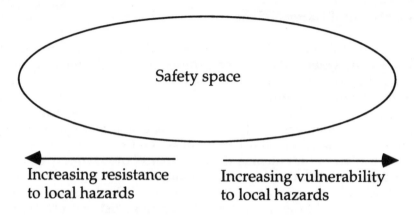

Figure 6.8.   The safety space

receive no kicks, a smallish number would be kicked once, an even smaller number twice, and only very few would sustain three or more kicks. Poisson's interest was in the people kicked, but let us consider the unkicked for a moment. Did they escape these natural hazards for the same reasons? It is highly unlikely. The chances are that some of them were plain lucky. Some were cautious when approaching the rear ends of their horses. Yet others were both cautious and had established a good relationship with their mounts.

Thus, although the Poisson distribution is usually represented as being one-sided, with a tail of increasing kick frequency going off to the right, it could – if the data were available – be converted to a two-tailed distribution with increasing degrees of resistance to being kicked going off to the left. Our interest is in the total 'safety space' covered by this two-sided distribution. A possible picture of this safety space is shown in Figure 6.8.

The space is drawn as cigar-shaped to indicate that while many people or organizations may occupy the central region, only very few will occupy the extreme positions of maximal possible resistance (which does not equal immunity) and maximal vulnerability. Absolute safety is unattainable. The value of the safety space is that it shows the only realistic goal of safety management: to get to the zone of maximum possible resistance and stay there. And it is the latter that poses the greatest challenge. After a serious event, most responsible organizations will institute repairs and reforms that will move them towards an area of greater resistance. But, in the absence of the appropriate measurements, it is hard to sustain such improvements, so standards relax and the organization drifts back towards the right-

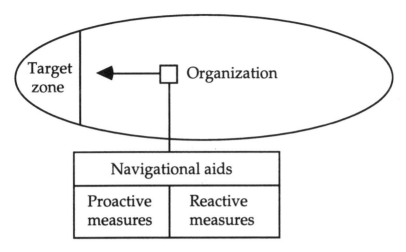

Figure 6.9. Aids for navigating the safety space

hand end of the space until re-energized by another accident, and so the to-ing and fro-ing goes on. The problem is that accident and event data do not provide an adequate account of where an organization lies within the safety space. These navigational aids can only be provided by a *combination* of proactive and reactive safety measures (see Figure 6.9).

Assessing organizational safety health

When a physician carries out a routine medical check, he or she samples the state of a few critical bodily systems: cardiovascular, pulmonary, excretory, neurological, and the like. From individual measures of blood pressure, ECG, cholesterol level, urinary contents, reflexes and the like, the doctor makes a professional judgement about the individual's general state of health.

There is no direct, definitive measure of a person's health. It is an emergent property inferred from a selection of physiological signs and lifestyle indicators. The same is also true for complex hazardous systems. Assessing an organization's current state of 'safety health', as in medicine, involves the regular and judicious sampling of a small subset of a potentially large number of indices. But what are the dimensions along which to assess organizational 'safety health'?

Principles of assessment

A variety of diagnostic approaches have been tried in applied research.

The individual labels for the dimensions assessed vary from industry to industry (oil exploration and production, tankers, helicopters, railway operations and aircraft engineering), but all of them have been guided by two principles. First, they try to include those organizational pathogens that have featured most conspicuously in well-documented accidents (i.e. hardware defects, incompatible goals, poor operating procedures, undermanning, high workload, inadequate training, etc.). Second, they seek to encompass a representative sampling of those core processes common to all technological organizations (i.e. design, build, operate, maintain, manage, communicate, etc.).

Since there is unlikely to be a single universal set of indicators for all types of hazardous operations, one way of communicating how safety health can be assessed is simply to list the organizational factors that we currently measure. Tripod-Delta, commissioned by Shell International and currently implemented in a number of its exploration and production operating companies, on Shell tankers, and on its contracted helicopters in the North Sea, assesses the quarterly or half-yearly state of 11 General Failure Types (GFTs) in specific workplaces: hardware, design, maintenance management, procedures, error-enforcing conditions, housekeeping, incompatible goals, organizational structure, communication, training and defences. A discussion of the rationale behind the selection and measurement of these GFTs can be found elsewhere (Hudson *et al.*, 1994).

Tripod uses tangible, dimension-related indicators as direct measures or 'symptoms' of the state of each of the 11 GFTs. These indicators are generated by task specialists and are assembled into checklists by a computer program (Delta) for each testing occasion. The nature of the indicators varies from activity to activity (i.e. drilling, seismic, transport, etc.) and from test to test. Examples of such indicators for design associated with an offshore platform are listed below. All questions have yes/no answers:

- Was this platform originally designed to be unmanned?
- Are shut-off valves fitted at a height of more than two metres?
- Is standard (Company) coding used for the pipes?
- Are there locations on this platform where the deck and walkways differ in height?
- Have there been more than two unscheduled maintenance jobs over the past week?
- Are there any bad smells from the low pressure vent system?

REVIEW, originally commissioned by British Rail (but now owned by Railtrack), measures 16 Railway Problem Factors (Reason, 1993a, b).

These are tools and equipment, materials, supervision, working environment, staff attitudes, housekeeping, contractors, design, staff communication, departmental communication, staffing and rostering, training, planning, rules, management and maintenance. (Note: The ten RPFs discussed in the context of RAIT were a subset of this complete list.)

REVIEW (and MESH – to be described in detail below) employ direct five-point ratings of the dimensions with regard to specific locations and tasks. The judgements, in each case, are made regularly be a large number of people in differing grades and occupations, via a computer program.

The third and most recent of the theory-based proactive instruments is MESH (Managing Engineering Safety Health), currently being used by British Airways Engineering. Since this is the instrument that has the most direct application to aviation, it will be discussed in some detail below.

MESH

MESH is a set of diagnostic instruments for making visible, within a particular engineering location, the situational and organizational factors most likely to contribute to human factors problems (Reason, 1994). Collectively, these measures are designed to give an indication of the system's state of safety (and quality) health, both at the local workplace level and in general. For convenience, the whole diagnostic package is implemented within a linked suite of computer programs.

Assumptions The basic assumptions of MESH are listed below:

- High standards of safety, quality and productivity are all direct functions of organizational 'health' – that is, the system's intrinsic resistance to accident-producing factors.
- Health is something that emerges from the interplay of several factors at both the local workplace level and the organizational level.
- A system's state of health can only be assessed and controlled through the regular measurement of a limited number of these local and organizational factors.
- MESH is designed to provide the measurement necessary to sustain a long-term fitness programme.

Accident-producing factors in the workplace Accident-producing factors in the workplace fall into three basic groups: human fallibility,

161

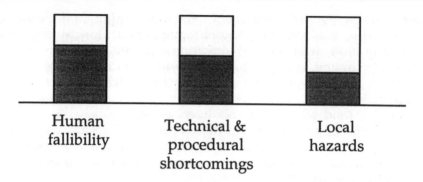

Human fallibility   Technical & procedural shortcomings   Local hazards

Figure 6.10.   Accident-producing factors in the workplace

technical and procedural shortcomings and local hazards. We can think of them as three buckets (see Figure 6.10).

The contents of each of these buckets can vary from time to time, but they will never be completely empty. Imagine that each bucket gives off particles. The fuller the bucket, the more it gives off. Let us also assume that accidents and incidents arise when these various particles combine by chance in the presence of some weak or absent defence.

MESH is designed to give an up-to-date indication of the fullness of the buckets. It does this by sampling selected ingredients in each bucket.

Assessing the local factors   Exactly what local factors are assessed depends upon the workplace. Some factors will be common to all locations, but others will vary from place to place. The following 12 local factors are measured in an operational hangar:

- Knowledge, skills and experience
- Morale
- Tools, equipment and parts
- Support
- Fatigue
- Pressure
- Time of day
- The environment
- Computers
- Paperwork, manuals and procedures
- Inconvenience
- Personnel safety features.

Assessments are made though simple subjective ratings of the extent to which each one of these local factors has been a problem in relation to a small number of jobs, days or tasks (how these are specified depends upon the particular work location). The assessments are made directly on a PC, using either the mouse cursor or the keyboard.

Ideally, these local factor assessments should be made by between 20–30% of the 'hands on' workplace in any given location. The assessors should be selected randomly, and each make weekly ratings for a limited period, say a quarter, after which a new set of assessors is randomly assigned, and so on. This method of selection is in keeping with the fact that MESH is a sampling tool.

Assessors are anonymous. On logging on to the MESH program, they are asked to give their grade, trade and location. On completing their ratings, assessors are provided with a profiled summary of their own input together with a cumulated profile for all ratings made over the past four weeks.

Figure 6.11 shows a schematic local factor profile. The *y*-axis indicates, in ordinal terms, the relative extent to which each of the local factors has constituted a problem in carrying out a variety of tasks over a given time period.

The profile shows, at a glance, which local factors are most in need of remediation. The intention is that the local management should focus on the worst two or three local factors only. Resources are always limited. It is much better to target specific problems than to try to tackle all of them. Thus, the local factor profile allows management to prioritize their quality and safety goals.

Assessing organizational factors   The same eight organizational factors are measured in each workplace. These are listed below:

- Organizational structure
- People management
- Provision and quality of tools and equipment
- Training and selection
- Commercial and operational pressures
- Planning and scheduling
- Maintenance of buildings and equipment
- Communication.

As with the local factors, assessments are made on five-point ratings scales in relation to specific jobs or tasks.

Organizational factors need to be assessed by technical management grades, people on the interface between the organziation at large and

163

Extent of
problem
y

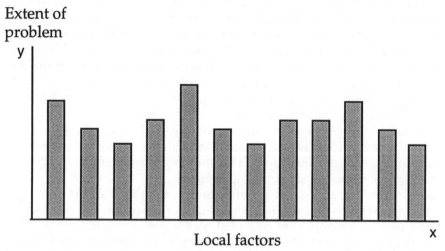

Local factors

x

Figure 6.11. A local factor profile (normally, the specific local factors would be labelled on the x-axis)

their specific workplace. Since organizational factors are likely to change far more slowly than local factors, assessments can be made at more infrequent intervals, say monthly or even quarterly.

A final word on proactive 'health' checks

It should be noted that relatively few of the organizational and managerial factors listed above are specific to safety. Rather, they relate to the quality of the overall system. As such, they can also be employed to gauge proactively the likelihood of negative outcomes other than coming into damaging contact with physical hazards, such as loss of market share, bankruptcy, and liability to criminal prosecution or civil law suits.

In all three of the diagnostic instruments, the measurements are summarized as bar graph profiles. Their purpose is to identify the two or three factors most in need of remediation and to track changes over time. Maintaining adequate safety health is thus comparable to a long-term fitness programme in which the focus of remedial efforts switches from dimension to dimension as previously salient factors improve and new ones come into prominence. Like life, effective safety management is 'one damn thing after another'.

Summary and conclusions

Striving for the best attainable level of intrinsic resistance to operational hazards is like fighting a guerilla war. One can expect no absolute victories. There are no 'Waterloos' in the safety war.

Effective safety management requires both reactive and proactive information in order to guide an organization to that region of the 'safety space' associated with the greatest resilience to operational hazards. In both cases, though, it is necessary to identify the organizational and situational factors contributing to unsafe acts. Such acts are like mosquitoes. They are best managed by draining the latent failure 'swamps' in which they breed.

References

Hudson, P.T.W., Reason, J., Wagenaar, W., Bentley, P., Primrose, M. and Visser, J. (1994) Tripod Delta: Proactive approach to enhanced safety. *Journal of Petroleum Technology*, **46**: 58–62.

Moshansky, V.P. (1992) *Commission of Inquiry into the Air Ontario Crash at Dryden, Ontario.* Ottawa: Ministry of Supply and Services Canada.

O'Leary, M. and Fisher, S. (1993) *British Airways Confidential Human Factors Reporting Programme: First Year Report (April 1992–March 1993).* Hatton Cross: British Airways Safety Services.

Pariès, J. (1994) Investigation probed root causes of CFIT accident involving a new-generation transport. *ICAO Journal*, **49**: 37–41.

Reason, J. (1993a) *REVIEW. Vol I. Management Overview.* London: British Railways Board.

Reason, J. (1993b) *REVIEW. Vol II. Theory.* London: British Railways Board.

Reason, J. (1994) Comprehensive error management in aircraft engineering. *Human Factors in Maintenance. Proceedings of the Aerotech 94 Congress.* London: Institution of Mechanical Engineers.

Index